I0486605

The Invisible Things of Him

Types and Shadows in the Creation

by

Austin D. Faulkner

© 2004 by Austin D. Faulkner. All rights reserved.

No part of this book may be reproduced, stored in a retrieval system, or transmitted by any means, electronic, mechanical, photocopying, recording, or otherwise, without written permission from the author.

First published by AuthorHouse 05/10/04

ISBN: 1-4184-6231-4 (e-book)
ISBN: 1-4184-3122-2 (Paperback)

Printed in the United States of America
Bloomington, IN

This book is printed on acid free paper.

TABLE OF CONTENTS

CHAPTER ONE

MANY MEMBERED BODY

Romans 1:20: **For the invisible things of him from the creation of the world are clearly seen, being understood by the things that are made.** John 1:3 says; **All things were made by him; and without him was not any thing made that was made even his eternal power and Godhead; so that men who don't believe are with out excuse.** This will be clearly seen in chapter 3 of this book. We know that there are many things in the old scriptures of the law and prophets that point to Jesus Christ and his first coming and also to his second coming. These things are often called "types and shadows". Moses delivered the people of Israel from Egypt and the bondage of Pharaoh. He was a "type and shadow" of Jesus Christ. Jesus has delivered us from the world system and the bondage of Satan into the kingdom of God. The children of Israel left Egypt with great riches. This is a "type and shadow" of the body of Christ being delivered from this world system into the Kingdom of God and its Glory in Him. What greater riches can a man have than eternal life and oneness with God? Corinthians 2:9 Paul says: **Eye hath not seen, nor ear heard, neither has entered the heart of man, the things which God has prepared for them that love him.** Many things

1

are found in the old scriptures concerning Jesus Christ. Are there no "types and shadows" of him in the natural creation? After all he is the one for whom and by whom it was created. The fact is, there are many things in the creation that point to him and to God's plan for mankind in him. In this book, we will cover some of these things in the natural creation that point to his purposes in Jesus Christ.

After the great flood of evolution and its accompanying philosophies of the beast, there was a huge rise in the "religion" of humanism and atheism. Men began to believe that they were the authors of their own destinies. Some even claimed that they were gods themselves. Others became atheists. They believed that the creation came about by chance without any Creator or any reason. Psalm 14:1 says: **The <u>fool</u> saith in his heart, there is no God. They are corrupt, they have done abominable works.** When we look at the universe and see the order in it, we certainly have the right to agree with this. The atheist is a fool! For instance, in chemistry there is the "Law of Combining Proportions" that says that atoms combine in simple ratios to form compounds. Two hydrogen atoms combine with one oxygen atom to form water.(H_2O) One carbon combines with two oxygen atoms to form the compound carbon dioxide.(CO_2) These simple ratios exist for all compounds found throughout the science of chemistry. Though each compound has its own distinct ratio, the ratio for each compound never changes. All compounds combine in the same simple ratios. If these combinations had come about haphazardly there would have been no order in chemistry and the science of chemistry would never have come about.

All the laws of physics have this same order. The laws of motion, inertia, thermodynamics, and nuclear physics are so ordered that scientists have been able to apply mathematical equations to them. In fact, if there had been no patterns and order present in the universe the science of mathematics would never have come about. Indeed, we would never have been sure of anything. Try to imagine the statistical chance that all this order came about by chance. It would more than likely take a computer 50 years to come up with the statistical number that it could happen by chance. Because of this mathematical order that God has placed in the creation, mankind has developed a great technology that threatens to destroy them. Scientists were very careful to set up standards in weights, measures and temperature so that their experiments could be preformed under the same conditions each time. Because of this, they were able to compare their results very carefully before drawing their conclusions and stating the laws of physics and other sciences. Any scientist who reported false results deliberately, lost his credibility with other scientists. We can truthfully say that our technology was based strictly on **TRUTH.** This is why our technological ability has far outstripped our social abilities. Because we adhere to truth in the sciences but not in our sociological relationships, we have become technical geniuses and sociological idiots. Jesus came and gave us the laws of truth such as those found in Matthew chapter 5 and 6 so that we would be able to develop a society of truth and peace but we pay no attention to them, in fact most of the world thinks they are foolishness. Jesus said, **"do unto others as you would have them do unto you."** If we did this there would be no murder because no one wants to be murdered. There would be no adultery, lying, false witnessing, or thief's because no one wants these things to happen to them. There would be no children without food or vaccines, because no one wants to see their children sick and hungry. There would be more concern about feeding the hungry and caring for the sick than there is about sending up space rockets. This is why I am so skeptical of the new "one world order" presently in existence. How can we conquer the universe and save ourselves, when we won't take care of each other? Do you see why we are so advanced technically but still have terrorism and war killing innocent people all over the world? Of course, the reason is that we adhere strictly to truth in technological things but every one has their own concept of truth in dealing with each other. **The ultimate outcome for selfishness, if given enough power, is self destruction.** God says this will happen if he doesn't intervene. Perhaps this is why we have these ultimate nuclear weapons; to prove that the words of Jesus in Matthew are truth. He said, my words will judge the world. Do you believe that a one world government will save mankind from itself?

Jesus came to give mankind the laws of truth and peace but only a very few hear and obey. The 8th chapter of proverbs verse 22 starts t*his way;* **The LORD brought me forth as the first of his works, before his deeds of old; I was appointed from eternity, from the beginning, before the world began. When there were no oceans, I was given birth, when there were no springs abounding with water; before the mountains were settled in place, before the hills, I was given birth, before he made the earth or its fields or any of the dust of the world, I was there when he set the heavens in place, when he marked out the horizon on the face of the deep, when he established the clouds above and fixed securely the fountains of the deep, when he gave the sea its boundary so the waters would not overstep his command, and when he marked out the foundations of the earth. Then I was the craftsman at his side. I was filled with delight day after day, rejoicing always in his presence, rejoicing in his whole world and delighting in mankind. Now then, my sons, listen to me;** blessed are those who keep my ways. Who was this that was with God from the beginning? It was Jesus. You who have read the whole chapter may say, this chapter is talking about wisdom not Jesus. However go to Corinthians 1 :

5

24. Paul says: **But unto them that are called, both Jews and Greeks, Jesus Christ the power of God and the <u>WISDOM</u> of God.**

We see then that Jesus Christ was co-existent with God the Father from the very beginning of the creation. Jesus was not only with God from the beginning, as John said, he was God. How can something exist **with** God, and yet **be** God? Suppose I set two bottles of compressed, life giving oxygen in separate rooms. These rooms do not have any air in them but there is an open door in the wall separating them. One bottle of oxygen is in each room. If the oxygen is released from both bottles, it will mix by the process of diffusion. The oxygen in both bottles will become **<u>one</u>**. It is impossible to tell which molecule came from which bottle. The gas in both rooms have the same properties and the same concentration. They are one. This is a crude physical way of representing the **<u>oneness</u>** of God. In the beginning, God declared himself to be love. He did not have to declare himself to be anything because, **he is** that **he is**. However, he intended to become a Father and he was going to be a **loving** Father. I do not know the entire ramifications of this loving oneness but thanks to Jesus Christ I have all eternity to find out. I also know that Jesus spoke by the commandment of the Father and nothing he ever commanded us to do can bring harm to anyone. His commandments were to love one another.

Jesus extends all through time. From his works of creation to his walk in the garden with Adam, his warning to Noah, his promise to Abraham, his instructions to Moses, his prophesies to Jeremiah, and Daniel, and he became flesh and dwelled with mankind. **<u>God manifested himself in the flesh.</u>** This is the mystery of God, Ephesians chapter 1: 9. Yes, throughout the old testament the prophets spoke of one person, Jesus the word. After Jesus finished his work he spoke of other sons. Paul said in Colossians that Jesus was the **first** born of God. This means there are to be others who will be born into the spirit. God intends to have many more sons. Jesus told his disciples **"Just as I and the Father are one so you will be one with me".** If it is possible for the overcomer to be one with Jesus and Jesus is one with God, won't believers who enter the Body be one with Jesus and also be God? I believe that the great vision found in the 1st chapter of revelations is the manifested body of Jesus Christ. Revelations 1: 12-18 says: **And I turned to see the voice that spake with me. And being turned, I saw seven golden candlesticks; And in the midst of the seven candlesticks one like unto the Son of man, clothed with a garment down to the foot, and girt about the paps with a golden girdle. His head and his hairs were white like wool, as white as snow; and his eyes were as a flame of fire; <u>And his feet like unto fine brass, as if they burned in a furnace;</u> and his voice as the sound of many waters. And**

he had in his right hand seven stars: and out of his mouth went a sharp two edged sword: <u>and his countenance was as the sun shineth in his strength.</u> And when I saw him, I fell at his feet as dead. And he laid his right hand upon me, saying unto me, Fear not; I am the first and the last: I am he that liveth, and was dead; and, behold, I am alive for evermore, Amen; and have the keys of hell and of death. Please note; it is this great body of Jesus Christ that has the keys of Death and hell, not Peter. This proves that the church is not built on Peter, but on the revelation of Jesus Christ that the Father revealed to Peter. This great image is the symbolic many membered body of Christ. The wonderful head with the eyes of fire and the countenance that shined as the sun, in his strength, was Jesus Christ the head of the body. The rest of the body represents the many membered body of overcomers from the apostles to those who overcome the anti-Christ in the last age. The brass feet that look like they have just come out of a furnace are the Holy Ones who come out of the anti-Christ furnace of tribulation, tried by fire. You can find the type and shadow of this furnace in Daniel chapter 3, and we will cover it in a later chapter. These brass feet come into the perfection of Jesus Christ right here on this Earth. They are the ones who are alive and remain, and are caught up to him, when Jesus comes again. His voice sounded like a voice of **many waters**. And why not, He is made up of many members. Perhaps this is why today's evil society has joined Pharaoh and Herod in the murder of babies. The devil knows that both male and females will be in this Body. That's why he's not seeking to kill just the males as he did in Pharaoh's and Herod's time.

Jesus was the perfection of selflessness. A lamb was sacrificed every Year by the Israelites at the passover. This passover lamb had to be perfect, one without a blemish. It was symbolic of Jesus. It was the **blood** of Jesus Christ that has redeemed us from eternal death. Paul said, "I die daily". (1 corinthians 15:31) This is the cross of Jesus that is applied toward the sacrifice of **self**. As was said before, selfishness is the main cause for all the evil that goes on throughout the world. Didn't Jesus say that we are to take up our cross and follow him? This cross is the Christian sacrifice of self. To enter into the body of Jesus Christ, **self must die.** Jesus told his disciples, **"The gentiles have those that are over them and these have those who are captains over them. But it shall not be so among you. He who would be chief let him be servant."** **(Mark 10:43-44)** Even though he had all things with the father he put it all off and became a servant for our sakes. He tells us that we must put away **all** selfishness and become an humble servant. **We must submit our will to his.**

But why do we have to go through the painful sacrifice of **self**? In order to see the real reason, we must see the vision of the coming body of Christ. This body is already a reality, it has

9

been a reality since the foundation of the world, but it is not yet manifest in it's **glory**. That will take place when he appears again. The scriptures say: **<u>my people perish for lack of a vision.</u>** I will now show you a vision of the body of Jesus Christ that is found in the scriptures. This may surprise you, but it is the image of God that is spoken of in the book of Genesis, chapter 1. This vision of the image of God is the first of the **"HIDDEN THINGS"** of the natural creation that we will cover in this book.

The real vision begins way back in genesis chapter 1; verse 26, it says: **And God said let <u>US</u> make man after <u>OUR</u> image and after <u>OUR</u> likeness.** Notice that God is speaking in the plural, **"our"** and **"us".** I believe he was talking to his beloved son Jesus.

But more than that, I believe he was talking to the entire body of Jesus Christ. Those in the body of Christ are not bound by time. What is your idea of what this image is? Is it the way of man? Mankind has been one of the most violent and evil creatures in this earth. Certainly this is not what he meant. Some people think that it's man's intellect. Even with his intellect he is unable to save himself because of his selfishness. That's why pollution is destroying the earth and we live in fear of nuclear annihilation. Even now mankind is fulfilling God's word in an attempt to save himself by the means of a new world order. Revelations 12: 11-18: By the use of this "one world order" he hopes to control the use of nuclear weapons. Just as you found the image of the many membered body of Christ in revelations, you can also find the image of this one world order of the beast in the 2nd chapter of Daniel. Some say that the image of god lies in man's ability to create. We Know that man can only create using the matter that God has already created. We know that he cannot create matter because one of the laws of physics says that matter cannot be created or destroyed. This is true except for a very few cases where unstable nuclear elements can be used to destroy matter. Man cannot create a new universe with eternal life inherent in it because he cannot visualize such an abstract universe as this would be. The universe he is in now is a universe of increasing entropy....DEATH! Many people deny the existence of God because they cannot visualize or explain him. No matter what he does to try to save himself he is still in a universe of death and Jesus Christ is the only door out of it. In order to see how man is in the image of God, we have to go to the 12th chapter of Corinthians beginning with the 12th verse. The apostle Paul received a wonderful vision which explains exactly what this image of God is. This word in Genesis was only the first part of the revelation of man's image, the rest of the revelation is given here in Corinthians. First, let me give you the dictionary definition of the word image: **IMAGE** (n): **a representation of a person or an object.** Something about man is a representation (image) of God. You know that God is perfect and any thing that

11

represents him must also be perfect. It's hard to imagine any thing about man that can fit the description of perfection, isn't it? Let's see what the Holy Ghost showed the apostle Paul about the image of God. Verse 12: **For as the body is one and has many members and all the members of that one body being many, so also is Christ.** Paul is talking about this body that we walk around in. **For by one spirit are we all baptized into <u>one body</u>. Whether we be Jews or gentiles, bound or free, we have all been made to drink of one spirit. For the body is not <u>one member but many.</u> If the foot shall say, because I am not the hand, I am not of the body; is it therefore, not of the body? If the ear shall say because I am not the eye, I am not of the body; Is it not of the body? If the whole body were an eye, where would be the hearing? If the whole were hearing where would be the smelling? Nay; but God has set the members every one of them in the body as it has pleased him. Now if they all would be one member where would be the body? But now they are many members but <u>one</u> body**. The eye cannot say to the hand I have no need of you. Nor the head to the feet I have no need of you. Nay, much more these members of the body which seem to be less important we bestow more abundant honor. And our uncomly parts have more comeliness. For our comely parts have no need: but God hath tempered the body together, having given more abundant honor to that part which lacked.

12

And whether one member suffer, all the members suffer with it; or one member be honored all the members rejoice with it. That there should be no division (discord), but that the members should have the same care one for another.

Speaking of the less comely body parts, consider the **immune system.** We have seen the outcome of what happens to a body without the immune system in this generation. The disease AIDS causes the body to waste away and eventuality die due to some disease it is unable to fight off. Consider the **wound healing mechanisms.** When the body is wounded histamine is sent to the wound site where it opens the arterioles to allow white cells to enter the wound to fight invading germs. Histamine causes the nerves at the wound site to become hypersensitive. Thus, the wound becomes very sore. This helps to prevent the re-injury of the wound. The **hormone systems** regulate the different functions of the body from sex to growth. Until this generation we hardly knew these beautiful (comely) body parts existed. They have been unknown down through the ages but now we know how important they are. Without them the body could not exist. Do you see how we are made in the image of God? Our natural bodies are made in the image of the spiritual family of God. The great body of Jesus Christ. Someday, perhaps very soon, this body will be manifested in all of the glory of God. In that day the body of Christ will be one with God. Yes, because of the mercy and love of God, **it too will**

be God! This may be a hard saying for you but it is true according to the Holy Scriptures. Romans 8:17 says we are **joint** heirs with Christ. This supernatural body will exist in complete love and unselfishness just like our physical body operates.

In order to understand what Paul is saying here, think about how your body operates. Each member has it's function and it preforms it's job in complete unselfishness and love for the body as a whole. What would happen if the eye became self centered? Suppose one day it said, "I think my great abilities are overlooked in this body. I'm going to set out on my own to find my real identity and individuality, in order to enhance my **self esteem.** If the body were driving down the highway it would probably hit a telephone pole. If it were walking in the woods, it might walk over a cliff. This is what happens to society when self esteem is prized more highly than the word of God. Our eyes, ears, hands, feet, heart, liver, immune system, and head are all many members yet **one** great unit. For example, suppose you were in the batters box in a baseball game. Suddenly the pitcher threw the ball right at your head. First you drop the bat, your arms go up to protect your head. Your legs buckle and you fall to the ground. The ball goes flying over your head. All these things happen in an instant. These members all act in **complete unselfishness** under direction of the head to protect the body. Such is going to be the great body of Jesus Christ. Isn't this a wonderful vision? Our natural bodies are made in the image of the family of God. Jesus and the Father are one and the body of Christ will be one with them. All one spirit and one mind. In this great unified body there will be no selfishness. It will be joined in perfect love one member for another. Wouldn't it be wonderful if the whole human race were joined like this? Every thing that is done would be for the common good instead of for selfishness and greed. Too bad some of our "freedom" organizations don't see that freedom should be for the common good instead of for self gratification. Without the love of Jesus Christ in our hearts, this will never be.

How does God look upon this body? He calls it Jesus. God knows it by its head. He will no longer remember its past sins. The blood of Jesus has covered them. The cross of Christ has made us perfect. We have a new name, *Jesus.* How can all this be? I'll give you an example. Suppose I put you in a group of ordinary people Then I cover every one's head so that no part of the head can be seen. I bring in your best friend and I tell him to pick you out of the group. Even though he is your best friend and has known you for a long time, it's unlikely that he can pick you out. But now I tell every one to take the covers off their heads. Your friend would know you right away. How do people know each other? Do they look at someone's hand or their belly or their neck to tell who they are? No. They look at their head. We are known by our heads. This is how God looks at his sons. They are known by their head, Jesus. How wonderfully intelligent God is to have made such a **perfect representation** of the manifested Sons of God. Think of how loving and unselfishly the head takes care of the body. The body loves and obeys the head without question. Though they are many members they are ONE body. God even now patiently awaits this precious fruit of the earth. (James 5: 7) Can you see why we must let the cross destroy **self.** Self has no place in this God body. Self esteem must not be prized so highly as our blind church psychologists tell us. There can be no self in the body of Christ. Indeed, there will be no need for it. God is complete and entire and has no lack whatever. There is no need for individuality or self in Him. God is love and Love shares all that he is with his beloved Sons. Whenever self enters in there is a division just like Paul says. Note the word in-**divid**-ual has the word **divide** inherent in it. The devil brought about a division in the sons of God when he exerted self pride. How terrible the consequences to humanity have been.

In God there is no lack. He is all and in all. In-<u>divid</u>-uality which is a form of exerting self is not needed. In God there is no place higher to go. **You are there! This is the wonderful fulfillment of the promise of God. This is the high calling of God in Christ Jesus.** Let the beast speak of self esteem. To go after his things is to strive for death. He will set himself up as God and speak of race purity and genocide to make a perfect genetic man. His talk is based on evolution which is the gospel of a jungle beast. You must stand fast in this evil day. Remember, your high calling in Christ Jesus. Hold fast to him. He has promised to never leave you nor forsake you. Tribulation is only for a moment when compared to eternity.

In this chapter we have seen that both Jesus Christ and the Son Body will be manifested at his coming. It is made up of many members, yet they are one Body. According to the revelation that Brother Paul had in Corinthians chapter. 12 they will be knitted together as one unit and they will have unbounded love for one another and for their head, Jesus Christ. **<u>Our human body is its earthly representation</u>**. (IMAGE) Think of how loving your body operates to protect the whole body while in complete submission to the head. This is a perfect representation of God's finished creation, the body of Jesus Christ. They will speak with the voice of many waters and will be of one mind. **In that day**, he said, **I will turn to the peoples a PURE language, that they may all call upon the name of the Lord, to serve him with one MIND.** Isn't this a wonderful vision of the high calling of God in Christ Jesus? We have seen that the human body is one of the **<u>hidden</u> <u>things that are made</u>**. It represents Jesus Christ and his body of Sons, the image of God. In the following chapters we will see other natural things that represent Jesus and his body.

CHAPTER TWO

GRAINS OF CORN

Romans chapter 1 verse 20says: *For the invisible things from the creation of the world are clearly seen, being understood by the things that are made, even his eternal power and Godhead so they are without excuse.* Paul is saying that nature itself shows things pertaining to Jesus Christ that were built into the creation. Even though the old scriptures speak of the things concerning Jesus. Nature also shows many of the things that God has purposed in Christ Jesus.

The whole creation was made for and by Jesus Christ, the word. Mark 4:28 says: *The earth brings forth fruit of itself.* (We will learn more about this statement in the last chapter of this book.) *First comes the blade, then the ear, then the full corn in the ear.* Many of you who read this have grown corn in the garden. Have you noticed the little blade that forms about half way up the stalk about where the stalk and a leaf join? This happens about midway through the growing season, in June or July here in the Ohio Valley. This little blade is the **"forerunner"** of the ear. The ear will form inside this blade. There are **THREE** stages of development of an ear of corn. Take note of this number three, it is a very important number. You will begin to see it's importance as you progress through this book. In the beginning, the blade gives the appearance of a new leaf. But in a few weeks, the ear begins to take shape inside of it. The blade, ear, and the grains symbolize the three dispensations of God's creation.

Please note that this blade comes from the stalk and always remains a part of it. When the stalk dies the blade and the ear die with it. Only the **grains** remain. This blade, "forerunner", becomes the shuck. It is the shuck that covers and protects the ear while it is developing. It remains with the stalk and falls back to the earth in death. Jesus is symbolic of the ear. Though he brought fourth the grains, he was part of the old law and he volunteered to die that the grains might live again. Though we know that Jesus has ascended to the Father to be our great high priest he had to die to bring fourth the grains. We find that the blade is "called out" to be the forerunner of the ear. Abraham was called out of the world ,by God, for a special purpose. He lived in pagan Babylon and was an idol worshiper. His purpose was to bring fourth the **Holy Seed.** God made a covenant with him for his **seed** to bless all nations of the earth. Just like the little blade, Abraham, Moses, The Law, and the prophets are all part of the scriptural "forerunner" of Jesus Christ. The special purpose of the blade is to bring fourth the ear and the seed in the ear. The Law and the prophets were given by almighty God to be the forerunner of Jesus Christ. The stalk comes of the earth and returns to the earth in death. It symbolizes the natural creation, as well as the Law. Before Jesus came all was death. Romans 5: 12,13,14 says: **Wherefore, as by one man sin entered into the world and death entered because of the sin of Adam, for all have sinned. For until the law sin was in the world: but sin is not imputed when there is no law. Nevertheless, death reigned from Adam to Moses, even over them that had not sinned after the likeness of Adam's transgression, who is the prototype of Jesus that was to come.** This is saying that from Adam to Moses there was no commandment (Law). Adam's sin was a transgression of the **commandment** not to eat of the tree, but, from Adam to Moses there was no commandment, therefore, people had no sin imputed to them until the law came. Paul says; (Romans 7:9) **T*he law came and I died.*** Paul tried to keep the law but found out that it was impossible, and cried out in Romans:7, ***Who will deliver me from the body of this death?*** He saw that every thing under the law was death because of sin. No man except Jesus has ever kept the law. We had to have a savior to deliver us. When any one encounters the law they cry out in helplessness, I need a savior. Buddha encountered the law of sin and death and cried out "NOTHING ALL IS NOTHING". Solomon encountered death also for he said, "I perceive that there is one event that comes to all men, both the wise and the fool." He cried out VANITY, ALL IS VANITY UNDER THE SUN. The blade is symbolic of the law which points out to flesh that it is of death and is in need of a savior. The blade, like this flesh, which is SELF, is under the law and must fall to the ground and die. But, praise God, the grains in the ear have the seed of life in them and will come out of the ground in a new body. Thanks to the ear, Jesus Christ, these are the ones who bear their cross daily until "old man SELF" is crucified with Christ. Then ,they live forever with him.

Jehovah made a seed covenant with Abraham. He told him that his anointed one was going to come through his seed. This promised **one** was going to bless all the families on the earth. (Genesis 17:7) Of course this one, that was to come, was Jesus Christ. All the prophets from Amos to Zechariah were of Abraham's seed. This seed covenant made with Abraham was not the only one that God made. God made a second one with a man named Phinehas, a Levite, who was the grandson of Aaron. This account is found in the book of Numbers chapter 25 vs. 11-15. This covenant was for the establishment of the everlasting priesthood through the seed of Phinehas. Jesus's mother Mary was a Levite of the daughters of Aaron. She was the cousin of Elizabeth, the mother of John the Baptist. The third seed covenant was made with David the King of Israel. This covenant established an everlasting kingdom through the seed of David. God revealed this to Nathan the prophet. He said the covenant would be fulfilled through David's sons. This account is found in 2nd Chronicles chapter. 17. These seed covenants were

word and <u>could not be broken.</u> Like the little blade they were the forerunner of the promised seed that was to come.

These people of the old scriptures were "school master's" and examples to this generation of Grace. Their experiences and tribulations were for the benefit of this generation. We were shown how they murmured and fell away into idol worship on the desert with Moses. Even after they came into the land that God promised them they continued to sin. In the book of Judges, when they began to sin the surrounding nations would send soldiers against them and overcome them. Then they would cry out to God for deliverance and he would forgive them and deliver them from their enemies. However, God's patience came to an end and he finally pronounced judgment on them. We all know how the Jew has suffered down through the ages because of the judgment of God on his sins. This is meant to be an example to God's people today. We also saw how God rewarded them with plenty while they were obedient to his commandments. Their land produced, the people were happy and healthy and their country was rich. But when they fell away into the worship of other gods they lost it all and were scattered throughout the world. Psalm 19 vs 9-12 says: ***The judgments of the Lord are true and righteous altogether. More to be desired are they than gold, yea, much fine gold; sweeter also then honey. <u>Moreover by them is thy servant warned: and in keeping them is much reward.</u>*** We have seen

23

how our country was founded by men who wanted the freedom to worship and honor almighty God. While this country kept His commandments it prospered and yielded its fruits. However, we are beginning to see the judgment of God upon it because we have turned away from God into paganism and the worship of material gadgets that are made by our hands. The idols that the Israelites worshipped were made by men's hands. Let his people of this day be warned by the schoolmaster, the blade.

These people were of the "old". Their purpose was to be an example to the people of the covenant and to bring fourth the **ear** Jesus Christ. Their experiences and tribulations were for our benefit. Anyone who thinks that the old scriptures are not necessary and are not needed by our new dispensation of grace are wrong. There are innumerable types, shadows and prophesies from which we learn the purposes of God. Paul. says they were our school master; we must learn from them.

The law shows sin for what it is, death! The law cannot be kept by flesh. It shows us that we are in need of a savior. The shuck (blade), the tassel, the whole stalk of corn, falls back to the earth in death. They are all part of the natural creation. But they point to a **new** creation that is yet to be manifest. They are born in a system of death. There is no way out of it, just as there was no way out from the law of sin and death before Jesus Christ opened the door to real life. This is why those who are still under the law (sinners) are so desperately in need of a **savior**. But, praise God, from the blade came the ear. It was brought fourth by that which is of death but the ear brings fourth **life. <u>The grains have life in themselves because they are formed in the ear not out on the stalk.</u>**

So we see that the blade represents the first dispensation of God's creation. We find the whole creation and the universe itself right along with the law and the prophets part of this first dispensation. The first member of the godhead, Jehovah, God the Father, was dominant in this time period. The people of this dispensation did not know him as Father. They did not see his great love for them as his **sons.** This would not be revealed until Jesus the **first** born son came into the world. Jesus, was God manifested in the flesh!

Sometime in midsummer the ear begins to form. Then, small rudimentary grains start to form **in** the ear. The corn silk forms at the tip of these grains, and then it grows from the grains down between the rows to the end of the ear. The tassel high above starts to rain pollen down on the corn silk below. (Note the type of the Holy Ghost which rains down on the newly converted Christian.) This little pollen grain of life begins to grow down the hollow silk right into the grains. Thus, the grains are fertilized, and life forms **within** them. At this point we must take special note of some things pertaining to the grains and the ear. First, the grains are

25

formed **only in the ear.** Second, they have **life within themselves.** It is the ear that gives them life. Third, the ear and the grains are **one body.**

The grains are formed **only** in the ear. Only in the ear can they ever receive nutrients and life. The devil has put a lie into many people's minds that they can come into Godhood without Jesus Christ. Some people think that they can become godlike by the process of evolution. Others think that salvation and godhood come about by a continuing reincarnation process of their flesh. Others think that their works will eventually make them gods after they die. There are others, worse yet, who think that they are already gods. These are the secular humanists. Just about any man who is honest within himself will admit to a belief that somehow or someway he will live after death . However, most of them do not believe the truth. The truth is that **only** in Christ Jesus is there eternal life. There is a "Babylonian" spirit of religion that believes that all religions will eventually bring about salvation. Remember in the book of Daniel that Nebuchadrezzer, the Babylonian king, placed Daniel in the same esteem as all the rest of his Priests. The Jewish God, Jehovah, was accepted along with all the other Gods. These people believe that Buddha, Krishna, Jesus and all the rest of the figures in religions of the world were "superstars" or "masters" sent by some unknown **force** that they call Karma. They believe this "force" is God. Daniel chapter 38 vs.37, speaks of the coming Antichrist: *And the king will do according to his will and he shall exalt himself, and magnify himself above every god, and shall speak marvellous things against the God of gods, and shall prosper until the indignation be accomplished; Neither shall he regard the God of his fathers nor the desire of women, nor regard any god; for he shall magnify himself above all. But in his estate shall he honor the god of forces and a god that his fathers knew not he will honor with gold and silver.* They do not believe that Jesus brought salvation. they say he was just one of the "masters"

But there was something different about Jesus. He not only raised from the dead as none of the others did, but he also ascended to God in the flesh. Jesus was not a risen spirit. He ate with the apostles. (John 21 vs. 9--; Luke 24 vs. 38- 43) Spirits don't eat fish breakfasts and you will never hear a spirit invite you to feel the wounds in his side and hands as Jesus invited Thomas, the doubting apostle, to do. This Jesus ascended to the **Father in the flesh.** He didn't disappear in the spirit to join some nebulous "force". In the seed covenants mentioned before, God promised these men that their seed, (Jesus) would stand before his face forever. He didn't say the *spirit* of your seed will stand before me. This is why Jesus, the seed of Abraham, David and Phinehas, ascended to the Father in the flesh. These **seed** covenants had to come to pass. Jesus was not just one of the "masters", he was **the** master.

These "grains" in Jesus, the ear, have life within themselves. Paul has said that those who sleep in Jesus Christ will someday hear his voice calling them to arise into eternal life. The 15th chapter of 1st Corinthians starting with the 22nd verse says: *For as in Adam all die, even so, in Christ shall all be made alive. But every man in his own order, Christ the first fruits (the ear) afterward those that are Christ's at his coming. (grains) Then comes the end when he will deliver up the kingdom to God.* The body of Christ, the overcomers, are the kingdom. *But some men will say, how are the dead raised up? With what body will they be raised up in? Thou fool, that which you sow (grain) is not quickened (made alive) except it die.* (Here he is talking about planting them in the ground. (the grave) *But that which is planted is not in the body that shall be, (come up in) but it is the bare grain. But God gives it a body as it pleases him. So is the resurrection of the dead. It is sewn in corruption; and it is raised in incorruption. It is sewn in dishonor it is raised in glory. It is sewn a natural body and raised a spiritual body.* Please read this in the 15th chapter of 1st Corinthians. The great apostle Paul likens those that are dead in Christ to grains. He is speaking of those Christians who are put into the grave in this corrupt flesh body but they are raised in an eternal spiritual body. What a beautiful vision! Who says that keeping the words that Jesus commands us to keep, isn't worth it? We have everything to gain and eternal death to loose.

Thus we see that the blade is symbolic of the first dispensation of God's creation. Are you beginning to see what the real eternal creation is? His intended creation is a body of sons fashioned into the image of his first born son Jesus. His sons will overcome the flesh and the world system. They will die to flesh but they will be born again into a spiritual body. They will all be one body with Jesus Christ , their head. The universe is only a huge school house. It is a system of eventual death. It is running down and slowly coming to a state of death. The stalk of corn, the law and the prophets, our bodies, are all of the first dispensation and are destined for death. The only purpose of the law was to show us the need for a savior. The law brings forth death because he that transgresses the law of God sins, and sin brings forth death. The flesh cannot keep the law no matter how much it tries, Eventually it cries out as the apostle Paul did, **Oh, *Wretched man that I am who will deliver me from the body of this death?*** (Romans chapter 7) However, praise God, Paul says we have the ear, Jesus Christ. He has made a new law that frees us from the old law of sin and death. Thus, the second dispensation lasted only 33 years, the life of Jesus. He has made the way for men to escape the law of sin and death. Jesus Christ is the second figure of the Godhead, **God in the flesh.**

However, most people prefer the old to the new. The devil is still telling mankind the same lie he told Eve, **"Oh, you shall not**

surely die.'' So men stay in the old way of sin instead of committing themselves to Jesus and the new man. Jesus said, ***no man who has tasted the old wine wants the new.*** That's the way it is between the flesh and the spirit. The old pleasures, the wealth and pride of life, the ways of lust and comfort are hard for most people to give up. The devil deceives them into believing that they can have salvation without giving up the world. But these ways all lead to death. The new way is sometimes hard. It is completely opposed to the world and it's lust. It is a life of denying self and being hated without a cause. You are misunderstood by your loved ones and those you work with. Those who live by all the words of Jesus and teach these things to others will also be hated by some of their own Christian brothers and sisters. These "church" people refuse to live by every word of God. But don't fret yourselves because of evil doers; we have life to gain and death to loose. Many people in the church say they are already born again. However Jesus said in John 3: 7 That those who are born again are like the wind, they can come and go, but you don''t know where they come from or where they go. Before the resurrection, when Jesus was still in the flesh, the scriptures say Jesus passed through their midst and went on his way. (Luke 4:30 also in John 8: 59). After the resurrection when he was really born again, the scriptures say that he vanished out of their sight. Or He appeared in their midst. (Luke 24:25-31; Luke 24 : 35-37; John 20:19and John 20 26). We saw a preview of the coming kingdom of God on the mount of transfiguration. This was to show us what it really means to be **born again.** The description of this preview of real life is found in Matthew chapter 16 with the last verse ending in the 17th chapter the 8th verse. **Verily I say unto you there be some standing here which shall not taste of death until they see the son of man coming in his kingdom, and after six days Jesus took Peter, James and John up into a high mountain apart and was transfigured before them: And his face did shine as the sun and his raiment was white as the light.** This is a brief preview of the glory that awaits those "grains" that overcome this world in Jesus Christ. The apostle John said we do not know what we shall be but when he appears we shall see him as he is.(now) (John 3:2)

As we said before, the 2nd dispensation was the birth, life and death of Jesus, It was only 33 years but **all** things that were spoken of him in the old scriptures were fulfilled during his life. He brought fourth the "grains" by the seed of the word. At his resurrection the "old" passed away, all things have become new. You have a choice to make. Will it be the old comfortable way of death or will it be the new, sometimes difficult way of self sacrifice in Jesus Christ.

The third dispensation is the maturing of the body (grains) of Jesus Christ. This is the last dispensation of almighty God's creation. His word will not return to him void. (without accomplishing

31

it's purpose) Jesus Christ, the Word has returned to the Father to intercede for the grains he brought fourth in the ear. The bible says that God patiently awaits the precious fruit of the earth. (James 5:7) He has sent fourth his Holy Ghost through Jesus's sacrifice to minister to the grains in the ear. The Holy Ghost is the 3rd manifestation of the **one** God. He ministers, comforts and teaches these overcomers. It is a narrow gate and a (hard) straight path that leads unto life and few there be that enter therein. Don't fret if you are lied on, your old life is thrown up to you, or you are called vile names, and persecuted for the name of Jesus. Remember the promise that was shown to Peter, James, and John on the mountain of transfiguration. You have glory to gain and death and hell to lose. **Praise the good word of God.**

In the day of the coming of Jesus, God will have a family made in his image. God has no lack he is complete! He has all power, all knowledge, he **IS** truth. Besides all this he is love. Because he is love he holds nothing back from his beloved **sons.** He shares all that he has with them. They become one. The blade then, represents the creation and the law and the prophets. God the Father is the main figure of the Godhead in this first dispensation of the creation. The ear that forms from the blade is the type of the second dispensation of the creation. Jesus Christ the second member of the Godhead is the dominant figure. The grains that form only in the ear and have life given to them by the ear are the type of the coming manifested sons of God. The Holy Ghost who ministers to these overcomers is the third and last member of the Godhead. If you see this vision of the coming manifested sons of God, give up SELF (selfishness) to the ONE who is the life. AMEN

CHAPTER THREE

THE NUMBER THREE

It was shown in the last chapter that God is going to finish his creation in **three** stages or dispensations. The maturing of the grains of corn is a type of the coming manifested sons of God at the return of Jesus Christ. We also stated that the number three was a very important number. It stands out as a <u>**type**</u> of God's purposes in Christ Jesus in the natural creation. Beside the stalk of corn, there are many other things in the natural creation that show God's plan to have a family of Sons with Jesus Christ as its head. In scripture the number three represents the just man made perfect in the body of Jesus Christ. It is God's <u>finished</u> creation. The three dimensional state of matter represents the <u>finished</u> state of all matter in the universe. The number three man that we will learn about represents the manifested Sons of the living God. The precious fruit of the earth that God awaits.(Jas 5:7) In the previous chapter we pointed out how the stalk, ear and grains pointed to the three dispensations of God's plan for mankind. In this chapter we will point out other things in nature that portray the three dispensations of God and also how they show the <u>three</u> members of the Godhead. Our awesome, infinitely intelligent God has made matter in such a way that it can only be explained mathematically using three dimensions. These three dimensions demonstrate the three members of the Godhead, and the three dispensations of the creation of the sons of God. Matter in its perfect symmetrical form also demonstrates the **oneness** of God even though there are three members in the Godhead.

As it was stated above matter can only be explained on the basis of three dimensions. All real objects have length, width, and depth. Why do we have to describe objects by numbers? Matthew 14:28 says: ***Whosoever doeth not bear his cross, and come after me, cannot be my disciple. For which of you, intending to build a tower, sitteth not down first and <u>counts the</u> <u>cost</u> whether he have sufficient to finish it?*** The reason is found right here in this scripture, we must express matter in terms of numbers in order to count the cost of building Something. Jesus was talking about the cross that we must bear in order to enter the kingdom of God. It's not a cross that we wear around our neck as a symbol. It is a cross we bear on our backs that burns out the impurity of sin and death and makes us into a house that God can dwell in. Jesus knew that men have to figure the cost of something they are going to build, that's why he gave this illustration. An engineer must also know exactly how much material to buy so he can purchase the correct amount of construction material to build whatever he is going to build.

The roofer deals with a flat two dimensional surface containing length and width. He must be able to figure how many roofing shingles to buy, otherwise, he might purchase too many shingles and he would have all his money tied up in shingles. On the other hand, if he didn't purchase enough, and it starts to rain late in the day before he finished, he would be in big trouble if the stores were closed. The concrete construction superintendent deals in length, width, and also depth. He must know exactly how much volume there is in the object he constructs in order to mix the correct amount of concrete. If he didn't, he would end up with too much. This would be a costly waste since it hardens up and cannot be reused. If he purchased too little, the concrete might harden before he could purchase more concrete to finish the job. So you see every one must sit down and count the cost whether they are building a spiritual house or a material one. All physical structures whether it be a box, ball, or any other shape must be figured mathematically on the basis of **three dimensions.** The volume of a ball is $4/3 \pi r^3$, where r is the radius. The volume of a box is S^3 , S being the sides of the box. All other structures have their own formula for figuring their volumes. The roof layer or the carpet layer is only concerned with two dimensions, length and width. I find it very fascinating to think that all matter throughout the entire universe has **3** dimensions. It is an amazing fact that no equation in our universe goes past the 3rd power, In our algebra classes we sometimes work with powers higher than 3 but if we tried to apply them to the real world we would find that after we pass the 3rd power equation, equations with higher powers are meaningless.

We are very fortunate that God made things in such wonderful order. The forces of Physics were made in such an orderly fashion that we have been able to apply mathematical laws to describe them. Everything in the entire universe was made in orderly patterns and symmetry. The human hand was made with exactly ten fingers. With them the first people who started to count could compare their results on the basis of the number 10. Imagine the confusion there would have been if some hands had seven fingers, others sixteen, while others had some other number. Our decimal number system is based on the number 10. All the peoples of the earth use the decimal system. How can these atheistic humanists look at the beautiful order in the universe and say; there is no God, it all happened by chance?

The formulas that we have mentioned are derived by a method of mathematics called, the calculus. I will attempt to describe to you what happens when these methods are applied without getting too technical. It is done by a special process of calculus called integration. Since most matter in the universe is in the shape of a ball, we will derive the formula for a ball. Its shape is called a sphere. I will also show how a Cube is created in order to show that all matter is mathematically derived the same way. God is a God of order. To start the process of integration we will start with the 1st dimension. This will be the length of the radius of the ball that we intend to create. The radius is the length measured from the center of the ball to its outer surface. Do you have a picture of the ball in your mind? Do you think that God had everything in his mind even before he spoke it into existence out of nothing? Let me ask you a question; can you picture anything in your mind with only one dimension? Our roof had 2 dimensions, length and width. What would a one dimensional roof look like? Our concrete block had 3 dimensions, length, width, and depth. Every thing has more than one dimension even the most minute hair. When we think about it we wonder if the first dimension really exists.

A calculus professor once told his class to think of the 1st dimension as an imaginary line or the **point source.** It seemingly doesn't exist yet it must exist in order to explain all matter in terms of numbers. Isn't this a wonderful example of the unseen God who is the **point source** of all things? The 1st dimension is a perfect representation of the 1st member of the Godhead, Jehovah, almighty God, the **point source** of all that was created. Like the 1st dimension, he doesn't exist; **or does he?** We can't see him, we can't prove that he exists, any more than we can prove a one dimensional object exists. But the very creation and the order in it, shows strong evidence that he **does exist.** Although we cannot see the 1st dimension alone, nor can we describe it, we know that it has to exist because the roof layer has to take into account both length and width when he figures the surface area of a roof. What a wonderful way for God to show us that he does exist.

How do we get to this 2nd dimension that has both length and width? As we said before, we will use a mathematical process called the calculus. The method of obtaining a plane area with 2 dimensions is called integration. If I am going to find the formula for a circle, the first thing to do is to start in the center at the "point source". Then, extend an **imaginary line** out from the center as far as the radius of the ball we intend to create. Sweep this line around the point source 360 degrees, one turn. We have inclosed a circle, haven't we? The space inside the circle is called the area. If we want to create a square object we sweep an imaginary, straight line, in a forward direction exactly the length of the side of the box. In the illustration below, it shows what happens by the integration process.

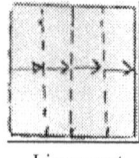

Line swept forward Encloseds the area Of a square

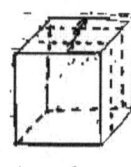

Area of square swept forward encloses a cube

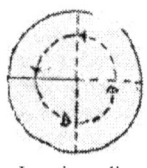

Imaginary line swept around encloses the area of a circle

Area (hoop) swept 360 to enclose the volume of a sphere

By preforming this operation we have a circle and a square. By the process of integration we find that the formula for the area of the circle is: $A = \pi r^2$. The formula for the area of the square is: $A = S^2$. We now have a two dimensional figure (length x width). It makes no difference how far we extend the radius of the circle it would still be a perfect circle. All matter that we see in the universe is in the form of a circle. Notice also that the formula has a square in it. (something multiplied by itself) Length and width is multiplied by itself. It is a flat surface just like a piece of paper. We now have something that is no longer imaginary. **It exists!** We can see it, touch it, and measure it. This dimensional plane is typical of the 2nd member of the Godhead, Jesus Christ. John said, **The word was God.** Paul said **God became flesh and dwelled with man.** Philip said, *Show us the father and Jesus answered, Have I been with you all this time and you haven't known me.*(John 14:8,9) **The pharisees said,** *You are not yet 50 years old and you say you have seen father Abraham? Jesus answered, Before Abraham was, I am.* **(John 8:58)** Jesus existed, his contemporaries saw him, touched him, he existed in the same flesh that they did. Jesus Christ was **God in the flesh!** He was not imaginary, he was real.

This two dimensional plane that we can see and touch is the type of Jesus Christ the 2nd person in the Godhead. God has made the second dimension of matter to be the type of his first begotten son, Jesus Christ. We saw in the previous chapter that it was the ear that brought fourth the grains, Without the ear there could have been no grain. The same holds true in this case. You cannot integrate directly from the 1st dimension to the 3rd. The 2nd dimension must be integrated first in order to bring about the 3rd dimension. Why didn't God just stop here? Why didn't he say the creation is finished? Jesus was perfect and he was without sin.. God had pronounced a death sentence on flesh that sins, but Jesus was not a sinner. God had given Jesus the choice to die or not to die for us. God said about him: ***"This is my beloved son in whom I am well pleased."*** Jesus said, __*No man takes my life I lay it down of myself.*__ As far as the Father was concerned **Jesus was complete in himself.** But Jesus knew from the very beginning, when the Father gave him the creation to preform, that the Father intended to have a family of sons. Jesus gave his life that we might be able to enter into the eternal creation **in him.** Without his unblemished sacrifice of self there would have been no way for sinful flesh to enter into eternal life. Jesus Christ then is the one portrayed in the second dimensional state of matter.

If you place the circle at eye level and turn it on its edge it disappears. Why? Because it has only 2 dimensions. One dimension of depth is still lacking. We must preform one more integration in order for matter to be complete as God created it. I will try to show you what happens when we integrate. Think of the circle this time as a hoop similar to the one that your mother used to embroidery designs on cloth. Take this hoop and set it on its end. Then turn the hoop 360 degrees (one full turn). It has integrated a space the same size and shape as the hoop. (See illustration above) This is the last step in the creation of the ball. It is now a sphere with the volume equal to $4/3 \ \pi \ r.^3$ The method of this integration process will not be explained here but please take my word, this is the formula for the sphere that is derived from it. Three dimensions, length, width, and depth are all integrated into the ball. The same holds true for the box illustrated above also. The mechanics of integration will allow us to keep right on integrating to obtain equations of 4th, 5th, or higher powers. But no equation higher than the 3rd power can be applied to the real world. Look in a table of formulas, if you have one. You will find that all formulas that apply to things in the science of physics or chemistry go only to the 3rd power, no higher. All matter in the universe has only 3 dimensions. This applies here on earth or on the farthest star. What an awesome and intelligent way that Almighty God has of showing his man the three members of the God head as well as the three dispensations of his creation of the body of Jesus Christ. And, since the universe is infinite, it also shows mankind the infiniteness of our God. We hear some speak of other deminsions or levels, but this is only a figment of their immagination. They are only thinking in the sense of deminsions because their minds cannot comprehend the realm

43

of spirit in which God dwells. God's work will be done in this 3 dinensional universe and there is only one door opened into the kingdom of God. That is Jesus Christ!

The plan of God's creation as shown to us in the bible consists of 3 stages. Some people call time allotments in the bible dispensations. The dominant figure in the Godhead that is shown by the 3rd dimension is the Holy Ghost. Jesus said to his disciples; *If I go away the comforter (Holy Ghost) will come. whom the Father will send in my name, and he will show you all things.* (John 14 : 26) Even though the Lord has gone to be with the Father his spirit is with us. He teaches and shows us the way wherein we should walk. The Holy Ghost is bringing into the kingdom all those who want to overcome the world. These are the seeds of Jesus Christ. **The true church.** The **real** creation that God purposed in the beginning will soon be finished. Jesus Christ is coming back to present the kingdom that he bought with his blood to the Father. By this time, I hope that you are beginning to see what the kingdom is. **It is the body of Jesus Christ, the sons of the living God!** This is the house that Jesus has built. When he said from the cross; *it is finished,* **he was talking about this house in which the spirit of almighty God will abide <u>forever</u>**. We don't yet know what we shall be but when he appears we will be like him. We saw a preview of the kingdom on the mount of transfiguration. Praise God; his word will not return to him without accomplishing its purpose. (Isaiah 55:11) His **word** (Jesus Christ) **has** returned to him, and he **finished God's work! <u>The word IS Jesus Christ and those *in him* are also going to be word in that day.</u>**

For those of you who may not believe in the Trinity of god **or the three members** of the Godhead, I would like to point out to you something about the finished ball that we have "created". It concerns the **oneness** of God. Take a good look at the ball. Can you tell which dimension is length, width, or depth? The ball is one unit isn't it? Look at a block that is a perfect square, can you tell which side is length or width or depth? If you take a perfectly square box and pick out the sides you want to be length or width or depth; all I have to do is flip the box over and you can't tell the ones you picked out. The ball, and the box are all **<u>one unit.</u>** This is true for any other 3 dimensional figure that is symmetrically perfect. The same holds true for the 2nd dimensional figure as well. We can't tell which is length or which is depth can we? Look at a circle, a perfect square. Which is width? Which is length? This signifies that God, the Father, and God, the Son, are one. There is no division or individuality in God. **HE is one God.** Though the 3 members of the Godhead are manifested as separate figures in each dispensation they are still **<u>one.</u>** And if you will accept it , it also shows that the finished creation, "the number three man" or the just man made perfect, is also one with God and Jesus Christ.

They too will be God. How awesome and intelligent is our God to have shown the three members of the Godhead in the very nature of matter. The <u>oneness</u> of God is represented also in the dimensions of a perfect circle, square, pyrimid or any other perfect figure.

I am going to show you some more things in *nature* that show the number 3 in the same way as the stalk of corn and the 3 dimensions of matter. These things show symbolically the 3 dispensations of God's creation of the body of Christ. First of all there are three primary tissues in the human embryo during the earliest stages of its development. The **first** and inmost tissue is the **endoderm**. This tissue goes to make up the digestive tract from the mouth to the anus. It is the inmost tissue in the body. The **2nd** tissue in the embryo is the **mesoderm**. Meso is the Latin word for middle. Mesoderm, therefore, is the middle part of the embryo. This tissue goes to make up the blood vessels, muscles and the internal organs such as the heart lungs and kidneys. Then the third tissue is the **ectoderm**. In Latin ecto means **outer**, and derm means **covering**. So ectoderm is the outer covering of the embryo. This tissue goes to make up the **"higher"**order of tissue in the human body. The brain, skin, nerves, sex gonads, endocrine glands and the pancreas are some of the tissues that are derived from ectoderm. This tissue is the control tissue. It regulates body functions, the immune system, and the reproductive mechanism. Can you see the **natural man** in the unsaved state as the endoderm, the **saved state** is typified as the mesoderm, the middle tissue, and the **spiritual man** in the higher ectoderm tissue?

The second thing that shows the number three in nature is matter itself. We have already shown that matter can only be mathematically demonstrated by the concept of three dimensions. But did you know the <u>all</u> matter can **exist** in 3 different states depending upon temperature and pressure. The whole universe is made up of matter in one or the other of these states. For example, iron is a solid under normal conditions. However, when heated to temperatures above 2000 degrees it turns into a liquid. If superheat is applied to iron it turns into a gas. Nitrogen under normal conditions is a gas. But when it is cooled to a certain temperature it turns to a liquid. If it is super cooled it becomes a solid. Mercury is a liquid under normal conditions but if it is cooled it "freezes into a solid. But if it is heated it will vaporize. **ALL MATTER** will react with heat, cold and pressure to exist in <u>**3** states</u>

Carbon is one of the most abundant of the chemical elements found in the human body. It is the "backbone" of compounds made by living things. Carbon can exist in three solid states. The first is charcoal. This is the rough porous primary state of carbon. It is like the 1st stage of the creation, the law and the prophets, or the natural unsaved state of man. Graphite is the next solid state. When charcoal is placed under increased heat and pressure it's rough charcoal state is changed into a flat planer state in the shape of a tetrahedron. These very flat structures slide quite easily over top of one another. This is why graphite is used as a solid lubricant. This is likened to the 2nd dispensation where Jesus Christ is the main figure of the Godhead. If placed under intense heat and pressure, carbon can be changed into a very hard translucent state. This state is the precious jewel diamond. The adverse conditions required to make diamonds is likened to the cross of Jesus Christ that purifies and makes his people perfect. Diamond typifies the "new man" the finished creation, the precious fruit of the earth that God patiently awaits. His precious jewels; the **son body that will be manifested at the coming of Jesus Christ.**

Water is one of the most common compounds found on Earth. This substance is one of the most essential things for the maintenance of life. In the scriptures water is symbolically used as God given spirit or life. For example, Moses smote the **rock** and water came fourth. The **rock** was a type of Jesus Christ being smitten for us on the cross that we might have eternal life. Our bodies are almost all water. The embryo is encased in water in it's mother's womb. Jesus said only those born of the water (water gushes fourth just before the baby is born) and of the spirit will enter the kingdom of God. Spirit beings that already exist will not be called the kingdom of God. Only flesh men are born of the water and the spirit. Water exists in 3 states. It's solid state is ice. (1st dispensation) It's liquid state is symbolic of the 2nd dispensation.

Vapor in it's gaseous state represents the 3rd dispensation. The three God given despensations for the creation of the perfect man.

The human brain is divided up into **three** separate areas according to function. The <u>hind brain</u> controls the autonomic functions such as breathing, regulation of heart beat, and smooth muscle function. An example of smooth muscles are the muscles that regulate the intestines for the elimination of waste. The <u>mid-brain</u> which controls the voluntary functions of the body such as walking and all other voluntary muscle functions that we do when we work or play. Most of these things are done without any conscious thinking. It would be very cumbersome to have to make a conscious effort to walk, wouldn't it? The <u>fore-brain</u> is in control of thinking, creativity, learning and decision making. The fore-brain can control all the other centers, to a certain extent. It can delay the elimination of waste, and the rate of breathing. It can in some cases, control the heart rate. Can you see how these three functions represent the Law, the grace of Jesus Christ, and then the finished creation as the body of Christ. The three dispensations of God.

Thus, we see that the number three is quite common in the natural world. God has stressed this to show to man his plan of redemption through Christ. He is saying; I am going to send out my word to do a work and this word will not return to me without doing what I have proposed for it to do. (Isaiah 55:11) Jesus Christ, **the word,** has returned to the Father and he has finished the work that it was appointed for him to do. God has shown us by the things that were created that his plan of salvation will be carried out in three stages. Three members of the Godhead have each dominated in His own dispensation. God the Father was the dominant figure in the creation and during the law and the prophets. God, the son, was here 33 years to preform His work and to finish it. God, the Holy Ghost, is here to minister to those who believe on the Lord Jesus Christ. Everything in the entire universe is made up of matter and this matter **consists** of three dimensions, whether it be a tiny atom or a huge star out in the universe. Matter can also **exist** in three states. Common things such as water and carbon that are found in all living tissue also exist in three states. Now let us see how God uses the number 3 in scriptures and what it means there.

We know that the 3rd dimension is the finished state of all matter. When the creation was completed all matter existed in a 3 dimensional state. We will find in the scriptures wherever the number 3 is applied it usually refers to the perfect, completed, spiritual man. However, it can also represent the three dispensations or the three members of the Godhead. This complete man are the ones who have overcome the world in Christ Jesus. They are the finished product of the work that Jesus preformed here on earth. The number three man is the just man made perfect; the precious fruit of the earth that God patiently awaits. (James 5:7) Jesus Christ was the first born of many brothers. When Jesus returns these brothers will put on immortality. The Mount of Transfiguration was a preview of this wonderful event. Paul says: (1st Corinthians 15:52) **In a moment, in the twinkle of the eye these overcomers who have come into perfection will be changed into the immortal sons of the living God.** At that moment, they will be born again. Many believe that we are already born again, but this is not true. We are conceived of the spirit, but not yet born into the spirit of God. The bible shows this very plainly. In Luke 4 :29 and John 8 :59 it says: "**Jesus passed through their midst and went on his way**"; when they meant to do him harm. This is while he was a flesh man just like the rest of us. However, after the resurrection in Luke 24 :31 "**And their eyes were opened and they knew him, and he vanished from their sight.**" Later in Luke 24 :36, "**And as they thus spake Jesus himself stood in the midst of them**". Thus we see that after the resurrection he could come and go like the wind. This is the definition he gave in John 3 :6 of those born of the spirit.

But when he appears again, we will no longer be children under tutors and governors (The Holy Spirit), but then, as full grown Sons, we will inherit all things. (Galatians 4:2) Peter says that we are born into a **lively hope.** (Peter 1:3) Peter says again in verse 22; that we are born again not of corruptible seed but incorruptible seed. Seed conceives, then that which is conceived must come to term before it is born. John says: **that man which is born of God cannot sin**. We know that all men sin until that day of the change into Glory. What a great vision! It surely makes us want to run the race to the end in order to win the prize. In that day all the brothers who are asleep in Jesus Christ will put on the same glory that he does. The third dimension finishes the creation of material things. The number three man symbolizes the finished creation of the spiritual man. This man, and his head, Jesus Christ is the sole purpose for the whole creation of God. This supernatural body is the only thing coming out of the natural creation that has eternal life inherent in itself. All the rest is like the stalk of corn it stays in the system of death. In that day, God's intended creation from the foundation of the world will be finished. The ungodly are not so but are like the chaff which the wind drives away.

One of the best examples of the number three in the scriptures pertains to the candlestick. It was to be placed in the ark of the covenant in the most holy place. Exodus 35:30 says: **And Moses said to the children of Israel, see. the Lord hath called by name, Bezaleel, the son of Uri, of the tribe of Juda and he has filled him with wisdom, understanding, knowledge, and all manner of workmanship; to devise curious works of gold, silver, and brass.** The types and shadows of the things made to be put into the arc of the covenant were so important that God picked out a man and endowed him with his own Spirit to become a clever craftsman. One of the things he made was the candlestick. The following is scripture pertaining to it. **Exodus. 25:30-32: Thou shalt make a candlestick of pure gold. Of <u>beaten</u> work shall the candlestick be made; his shaft, his branches, his bowls; his knops. and his flowers, shall be of the same. (pure gold) And <u>six</u> branches shall come out of the side of it;<u> three branches of the candlestick out of one side, and three branches out of the other side</u>. (Exodus. 25: 36) Their knops and their branches shall be of the same: all of it shall be <u>one beaten work of pure gold.</u> (Exodus. 30: 29) and Thou shall sanctify them that they be most holy. And whoever touches them must be holy.** The significance of this candlestick was to be considered most holy. Read on to see what a high calling God has for mankind in Christ Jesus!

There are many things in this small bit of scripture that is significant of Jesus Christ and those who are made perfect in him. First is the man Bezaleel who was sanctified (set apart) by God and given special power as a craftsman. In this case his work was with **one piece** of pure gold. The gold in the temple always signifies the glory of God. Just about every thing in the holy of holies; the place in the temple where God abides, is made of pure gold. Only that which is changed into or created of the glory of God can exist in his presence. This candlestick was made of one piece of gold. Not one bit of it was welded or melted on, it all was beaten from **one** piece. There was one central shaft and **six** branches. **Three** branches coming out of one side and **three** branches coming out of the other. The central shaft signifies Jesus Christ the Holy One of God. There are six branches. Six always signifies the number of a man. It is symbolic flesh. However you will note that <u>three</u> of these six branches come out of one side and <u>three</u> branches out of the other. This is symbolic of flesh that has been made perfect. The number three in the candlestick symbolizes the flesh man who has been made perfect **in** Christ Jesus. They are beaten out of the same lump of gold that the central shaft was beaten from. **They are one.** This is symbolic of the manifested Sons of God. Of course the central shaft, Jesus, is the HEAD. You see in this case we have seven candles. Seven is the number that symbolizes God. This **body of Christ** will be one with God, therefore, it will be God. These are the sons of God. They are the kings and priests of the kingdom of God. The six branches signify flesh, but they were split up into **three** branches. The number three symbolizes the manifested son, the ones who have truly been born again.

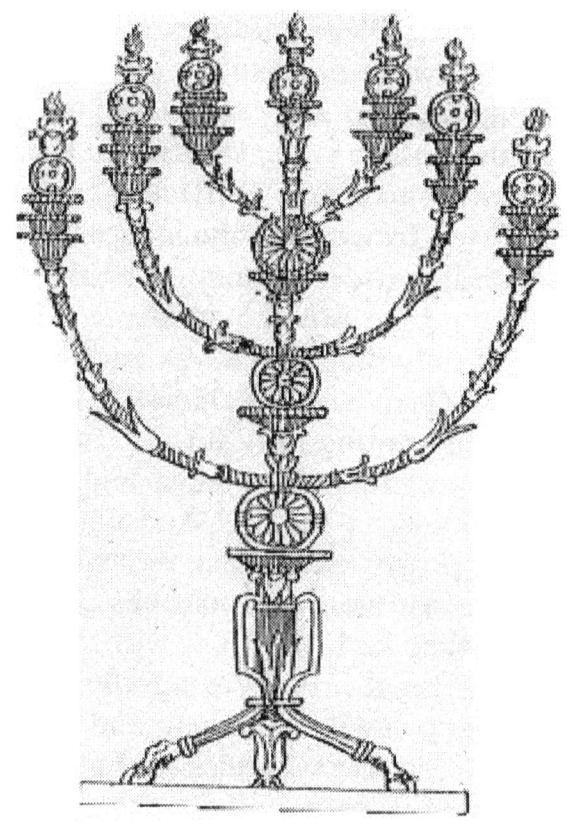

Above is an example of how the Candlestick looked

For the next example of the number three in the scriptures we will move to the book of Daniel. According to the prophesy of Jeramiah, (Jer 4: 2) God sent his holy remnant of Jews right into the kingdom of Babylon, the very seat of satan. This happened 30 years after the prophesy about the good basket of figs. And you can be sure that satan knew about it. They were delivered into Babylonian captivity. His children went right on worshiping

God the way they always had. Daniel, of course was one of the remnant that was sent with them. Daniel must have been a prophet even back in Israel because he was put in the same living quarters with all the other religious priests of the Babylonian kingdom. Daniel had **three** other Jewish companions with him. Their Jewish names were Hananiah, Mishael, and Azariah. Their names were later changed to Shadrach, Meshach, and Abednego by king Nebuchadrezzar. One night, God caused the Babylonian King to dream a dream concerning the end of time. However, the king forgot what the dream was all about. None of his astrologers and soothsayer could reveal or interpret the dream so the king ordered them all to be killed. When Daniel heard of it he prayed to God about it. God revealed to him the dream and also the interpretation of it. This so pleased the king that he acknowledged Daniel's God to be the most high God and made Daniel the 2nd in command in the Babylonian kingdom. In this way God was able to protect his Jewish remnant while in Babylon. The devil set about immediately to destroy this trust in God's people. Daniel 3: 1 says: **Nebuchaerezzar the king made an image of gold, whose height was <u>sixty</u> cubits high, and the width thereof was <u>six</u> cubits. He set it up in the plain of Dura, in the province of Babylon.** Please note the fact that the idol has the number **sixty six** in it. We will see the significance of this later on in this chapter. Daniel chapter 3, tells us that king Nebuchadrezzar sent out a decree that every one in the kingdom was to fall down and worship this golden image when they heard the sound of his evil music. If they did not, they would be thrown into a fiery furnace. The evil priests knew that Shadrach, Meshac, and Abednego refused to worship the image, and they told the king. This caused the king to be very angry, and he ordered them to be thrown into the furnace of fire. But much to the surprise of the king the fire had no effect upon them. He observed them walking around in it, but not only that, there was another person in the fire with them who had the likeness of the son of God. Then in Daniel 3; 2 it says: **Then Nebuchadnezzar came near to the mouth of the burning fiery furnace, and spoke, and said, Shadrach, Meshach and Abednego, Ye servants of the most high God come forth, and come hither. Then they came forth of the midst of the fire. And when they all gathered around them, they saw that the fire had no power over them, nor was the smell of fire on them. Not one hair on their head was singed nor were their clothes changed by the heat.** The only thing burned was their bonds that they had tied them with. We will see the significance of this later. This bit of scripture has great prophetic significance. **(Revelation 13: 11:18) And I beheld another beast coming up out of the earth; and he had two horns like a lamb, and he spoke as a dragon, and he does great wonders so that he makes fire come down from heaven on the earth in the sight of men, and deceives them** that dwell on the earth by the means of those miracles which he had power to do in the sight of the beast; saying to them that dwell on the earth, that they should make an image to the beast, which had the wound by a sword, and did live. And he had power to give life unto the beast, that the image of the beast should both speak, and cause that as many as would not worship the beast should be killed. And he caused all both small and great rich and poor, free and bond, to receive a mark in their right hand, or in their foreheads: and no man had the right to by or sell, save he had the mark, or the name of the beast, or the number of his name. Here is wisdom. Let him that has understanding count the number of the beast: for it is the number of a man, and his number is 666.

Note in the scripture from Daniel above, it said that Nebuchadrezzar's image was 60 cubits high and 6 cubits wide. Nebuchadrezzar's kingdom represented the number 66. However, the Antichrist kingdom represents the number 666. In the scriptures God says the number six is the number of a man. It is man in his unsaved state, or man controlled by satan. Since there were only two six's in Nebuchadrezzar's kingdom, it means that it had not come into the **completeness** of evil. The Antichrist kingdom has **three** six's. This means that it is the completeness of evil, run by a man with satan himself in full control. No known being is more evil than satan.

But this is not all that I want to point out in this. You will note that there were **three** men in the furnace of fire. They were thrown into the fire because they did not worship the idol that Nebuchadrezzar had made. These men represent the people of God who refuse to worship the image of the beast in the Antichrist kingdom. They will be thrown into a furnace of great tribulation. Some will be killed but some will come out of this tribulation completely without harm. The Lord will be with them during this tribulation period. the only thing that will be burned from them will be the bonds of death. They will be in the likeness of the glory of God. Free of death forever! These three men represent those who will come into the likeness of Jesus Christ right here on the earth. They are the body of Christ, the purpose of god's creation. These are the ones who will put the devil and his evil pagan kingdom under their **feet.** Note the difference in these feet and the feet of iron and clay that are in the Antichrist body in Daniel; chapter two. This **foot company** are the ones who will smite the feet of the Antichrist body in the future and destroy the Babylonian system forever. They are the part of the Body that is in contact with the earth. They are the ones changed into the Glory of God right here on this earth. Praise the holy name of Jesus! The scriptures above say that the feet are brass. The brass things in the arc of the covenant, are symbolic of flesh. They are usually found outside of the Holy of Holies. These brass feet look like they have

59

been refined in a furnace. Since brass represents flesh, it is safe to say that these are the people (flesh) in the Antichrist kingdom who go through a furnace of great tribulation. Like the three Jews in the furnace of fire who refused to worship Nebuchadrezzar's golden image, Jesus will be with these people. Some will be killed that they may be made white and clean. But some of them, will come out of this furnace of tribulation unharmed, Jesus will be with them just as he was with Shadrac, Meshac, and Abegnago. These will put on the glory of God right here on this earth. Paul says; **we will be changed in a moment in the twinkling of a eye.** These are the ones who will enter into the body of Christ without going through physical death. That's why they appear as **brass feet.** The bonds of death are gone forever.

Our next illustration of the significance of the number three in scriptures is found in Matthew 16:27 to 17:3; also Mark 9:2 has an account of it. Jesus had been teaching his disciples when he said; **Some of you sitting here shall not taste of death before you see the kingdom of God come in power: and after six days Jesus taketh Peter, James, and John his brother, and brought them up into a high mountain apart, and was transfigured before them: and his face did shine as the sun, and his raiment was white as the light. and, behold, there appeared unto them Moses and Elijah talking with him.** Note that Jesus took **three** men into the mountain with him. This **three** signifies the overcomers that

Jesus will take into the kingdom of god with him. These three men saw a preview of the kingdom of God before they were martyred. Jesus changed into the glorious body that he will come back to this earth in. **This is what being born again really means.** Peter later on in his epistle speaks about being in the Holy mountain with the Savior. (Peter 1:18) He speaks of being born again into a lively HOPE. (Peter 1:3) Do you see the great vision of what it will be like to be born again? No wonder the apostles faced death for the blessed name of Jesus. Death has no power over the people of God.

There are other things also that show the number three and its significance. Genesis 6:10 says: **Noah begat three sons Shem, Ham and Japheth.** Here we see the number three again. These men symbolized **a new beginning** upon the earth. Jesus likened the last days to the time of Noah for he said **as it was in the time of Noah so shall it be at the time of the coming of the son of man.** Noah's **three sons** symbolize the sons of god who will make a new beginning on the earth. It will be pure and clean after the antichrist kingdom is destroyed. At this time there will be a new beginning, old things shall pass away and all things will be made new.

Exodus 23:14-16; Exodus 34:18-22; Leviticus 23:6,34,39; These are scriptural references for the things I am about to discuss. They describe **three** feasts that God commanded the people of

Israel to have during the year. The first one was the feast of the passover. It was also called the Feast of Unleavened bread. This of course signified the death and resurrection of Jesus Christ. His blood was shed for us and is applied to the "door posts" of our hearts by believing on him for eternal life. The 2nd feast was exactly **50** days following the Feast of the Passover. (Leviticus 23: 16) This feast was called the feast of The First Fruits, or the Feast of Weeks. This feast signified the day of Pentecost which came exactly **50** days after Jesus's resurrection. Approximately fifteen hundred years after this feast was instituted by God The Holy Ghost fell upon the little group of apostles gathered together to celebrate it. They were the "first fruits" of the early church. Without them there would be no church today. Read chapter two of the book of acts. Note the number **50.** Whenever you see the number **5** in the symbolism of the scriptures it means **God's Grace.** The **third** feast instituted by God was the Feast of the In-gathering or it was later called the Feast of Tabernacles. It came at the end of the year at the time of the harvest. This feast signifies the events that will happen in the end times when the overcomers of the Antichrist and the martyrs for the name of Jesus will be gathered into the body of Jesus Christ. These are the precious Fruit of the Earth the God almighty awaits. These are the seed of Jesus Christ. Please remember that these Feasts were instituted many hundreds of years before the birth of Christ.

(Judges chapter 6,7:) This tells of a man named Gideon. God picked him to deliver the people of Israel from the nation of Midian. God used him to show that by his power, he could deliver the people of Israel from any situation and against any odds. Gideon started out with 32,000 men in his army. God told him to tell all those in his big army to go home if they were afraid in their hearts to go into battle. He lost 10,000 men because of this. The Lord said, this is still too many. Take them down to the water and I will **try** them there. Choose only those who lap the water up with their hands and drink. (Judges 7:6,7) Those who fall down on their bellies and drink like cattle with their mouth, send them home. This writer once heard a preacher say that those who drank from their hands were looking for the enemy up in the hills. They were concerned about the whole group, not just themselves. The ones who fell down and drank with their mouths were concerned only about themselves. They were selfish! Now take note. There were only **300** men that drank from their hands. God took these and divided them up into **three** companies and defeated an army of thousands the next day. I believe that this number three is significant of the end time army called, JOEL'S ARMY. This great army will rid the world of the Antichrist and his armies forever. (Joel 2: 2,3)

(Kings 7: 4) Elijah stretched himself **three** times over the widow's child to bring him back to life. The manifested Sons of

God are the ones Jesus was speaking of that will have the "double portion" to do the greater things.

(Matthew 13 : 33) Jesus spoke a parable: The kingdom of God is like unto leaven; which a woman took and hid in **three** measures of meal, till the whole was leavened. This indicates the three dispensations of the Godhead.

(Luke 10 :36) Three men passed by the man lying in the ditch in the parable of the good Samaritan. This man signified the human race beaten and wounded and left to die by satan because of Adam's sin. The first man was a priest. he signified the priesthood of the world religions of that day. He passed to the other side. The next one to pass him by was a Levite. He signified the priesthood of Israel. He also did nothing. Then came the good Samaritan (Jesus) , He ministered to him and then said he was going away (return to the Father) but that he would repay anything yet to be paid when he returned. Is this the payment of judgment on the wicked men and demons who have mistreated God's elect children down through the ages?

(Acts 10: 9-19) Peter sees a vision of all manner of unclean things descending from heaven on a sheet. This happened **three** times. This, of course signified that the gospel of Jesus Christ was to go to the Gentiles. Just after this **three** men knocked on the door who were sent from the house of Cornelius, a gentile Roman soldier.

(Zechariah 13: 7-9) Awake, O sword, against my shepherd, and against the man that is my fellow, saith the Lord of hosts: smite the shepherd, and the sheep shall be scattered: and I will turn mine hand upon the little ones. And it shall come to pass, that in all the land, saith the Lord, two parts therein shall be cut off and die; but the third shall be left therein. And I will bring the third part through the fire, and will refine them as silver is refined, and will try them as gold is tried: they shall call on my name, and I will hear them: I will say, It is my people: and they shall say, The Lord is my God. This scripture speaks for itself.

Genesis 18:2; Is another scriptural reference to the number three found in the bible: **Three** men came to destroy Sodom. I believe that it will be the manifested sons of the living God who will destroy the evil sodomite anti-Christ kingdom. Here again, these **three** men are symbolic of the just men made perfect who will put the anti-Christ kingdom under their feet. Thus we have seen the great significance of the number three in both nature and the scriptures. There are many other instances of the number three found in the scriptures. Read your bible see for yourself. Isn't this a wonderful vision? We certainly have reason to run the race to the end, don't we? Remember this vision when the power of hell comes against you. Stand!

CHAPTER FOUR

SUN, MOON, AND THE TREES

For the invisible things from the creation of the world are clearly seen, being understood by the things that are made even his eternal power and Godhead, so they are without excuses. We have already seen this scripture in the preceding chapters, but I would like to emphasize it again since this is what this book is all about.. It is important that we know that nature itself contains types and shadows of Jesus Christ, just as the scriptures do. We must realize that all the physical creation is under the law, therefore, it is under the law of sin and death. Only that which is in Christ Jesus has overcome the law of sin and death. Yes, the Lord put things in his creation to show his purposes. We have already seen several things in the previous chapters that are types and shadows of God' purposes in Christ Jesus. In this chapter we will see more things that point to his purposes that we encounter every day. There are also things in the creation that show the evil one who opposes God's plans for mankind as well. This evil being, Lucifer and his evil demons, are a reality that must be encountered by every Christian who sets his face to enter the Kingdom of God.

Lets begin with something that shows Jesus Christ as nothing on this earth can. Did you know that the sun is a type of Jesus

Christ? The sun has been worshipped by men all down through the ages. Roman sun worshippers entered the church of Jesus Christ and corrupted it for centuries. Even today it is worshipped in various ways. The secular humanists of today have a form of sun worship as the "life giver", which in a sense is true. However, the son of God, Jesus, is eternal. The sun is not, it is dying. Some day it will either explode into a supernova, or, it will become a red dwarf and eventually die as a black hole. When it dies all flesh must die with it. Have you ever watched a sun rise? As it rises in the east the shadows of darkness flee and the world is encompassed in light. John chapter 1 : 4 says: *In him was light and this light was the life of man.* Light shows both beauty and ugliness for what it is. The light of the sun brings out the beautiful colors of an autumn day and it also exposes the sordid ugliness of a garbage strewn back alley. Everything is laid bare by the light of the sun. The light of Jesus Christ will lay bare all the evil that has been done by evil men in the coming judgment. Evil people hate light. Their deeds are manifested in light, therefore, they hide them in darkness. They know that their evil intents and purposes will be laid bare in the light of truth. Most all carnivorous beasts of the jungle hide in the shadows of darkness to prey on their unsuspecting victims. The terrible carnage of war is spawned in the dark closets of greedy men and evil politicians.

The sun is a type of Jesus Christ the true source of life. The life that he gives is eternal. The sun gives life also but it is seasonal, and the life that it gives, sooner or later must die. In the spring, as the dark cold days of winter recede into the warm April sunlight, we see the little seed spring fourth from the earth in its **new body.** (see the 15th chapter of 1st Corinthians) This is nature's type and shadow of the newly saved Christian who has committed his life to Jesus Christ. He too, will someday be given a new body when the Lord appears. (1st Corinthians chapter 15). The little seedling raises its arms toward heaven as it grows in strength. It gives up water vapor that rises heavenward and falls back as the warm spring rains, these rains make the life within it more vibrant so that it can withstand the enemies of life that encompass it in a hostile world. The new Christian raises his arms in prayer. And their prayers rise heavenward like fine incense to God who causes the Holy Ghost "rain" to fall on them. He makes them strong and healthy in the spirit. This Holy Ghost rain enables the new life of Jesus Christ in them, to withstand the hostile demons of satan. They want to kill this God given life that now resides in them. A famous atheist once said: "If all the people in this world, who call themselves Christians, lived according to the words of Jesus Christ, this world would become a virtual utopia. For a brief moment the light of Jesus Christ struck the barren soul of this atheist, and life sprang fourth for a time, but it soon died in the cold icy winds of doubt and unbelief. This is similar to a seed on a barren, rocky crag of a high mountain top. The warm light of the July sun causes the life in it to spring fourth. A beautiful flower is born but in a few days it dies as the cold winds from the high mountain glacier kill its new life.

Some of you will say, "But I'm already alive, how does Jesus give me life"? Are you really sure that you're alive? We encounter a term in the study of thermodynamic systems that do work. This term is called entropy. Entropy can be explained thusly: 'It is the degradation of the matter and energy in the universe to the ultimate state of uniformity and inertness'. This is just a fancy way of saying that the universe is constantly loosing energy. This energy cannot be recovered; it is lost forever. The universe will eventually lie cold and still in an inert, uniform state of absolute zero..... **death!** This brings to mind the Word spoken in the scriptures; **The heavens and the earth shall wax old as a garment and be folded away but my word will never pass away. (** IS. 51 :6 and HEB. 1 : 11**)** Our sun is one of these thermodynamic systems that do work. The sun is constantly doing work on the earth to maintain its physical life. However, our sun is slowly dying. The heat it gives off to warm the earth is the form of energy spoken of above that we call entropy. Entropy is the energy that is lost as heat in the nuclear furnaces of the sun. This heat is lost forever, it cannot be recovered. The little man with the enlarged, bald, toothless head and the spindly legs; who is the product of the imagination of the evolutionist, the superman of the future, has no hope of eternal life. When we view him in the light of entropy, we see him lying cold and still, frozen forever in his self made frozen tomb, as the red sun of the earth blinks out into a magnetic mass called a "dark hole". But, the words of Jesus Christ, **"Know you not, that you must be born again to enter the kingdom of God"**, will live forever in the body of Christ.

Many years before Jesus Christ was born a young man named Siddhattah Gautama lived in India. He became obsessed with the reality of disease, poverty, old age, and death. He wanted to find out what life was all about. One day he saw one of the many wondering ascetics, who lived their lives under very strict rules and spent much time subduing their bodies in religious meditation. He decided to join them in order to seek wisdom and truth. So he left his wife and new born son and went into a life of asceticism and meditation. After several years of this life of self denial, he woke up to the fact that fasting, sleeplessness, and self torment was not the answer to finding wisdom. Later Gautama saw that the selfish desire for money, self gratification, and immortality were the evil things that brought about the misery in life. Since death was the only outlook for mankind, he said we should abandon these selfish desires and serve mankind. He ended up by coming to the conclusion that all was nirvana (**nothing**). These ideas he proclaimed about nirvana have been changed down through the ages, but many historians believe that he encountered the specter of death and proclaimed that all is nirvana. Gautama later became known as Buddha.

Along about the same time another man was born in the nation of Israel. He was also a seeker of wisdom. His name was Solomon. He too, set out on a quest for wisdom. His account is found in the book of Ecclesiastes in the bible. He called himself the preacher. He first tried the life of a playboy. He didn't deny himself any pleasure. But he soon found that this was not the answer to life. He said vanity, vanity, all is vanity under the sun and a striving after the wind. He then set out to be a great builder. He built great edifices. He built God a beautiful temple and himself a beautiful mansion. He made gardens filled with flowers and plants. It was so wonderful that people came from far and near to wonder at them. Again he ended up unsatisfied with his life. He then decided to live the life of self denial just as Buddha did. He denied his body of food and pleasure, and sat in sackcloth and ashes. He perceived that **<u>ONE</u>** event comes to all men both the wise man and the fool. <u>This event is death.</u> He saw that all his labor and endeavors came to nothing when he went to die. This again, made him cry out, **"vanity, vanity, all is vanity under the sun and a striving after the wind. All that is left for a man to do is love the Lord God with all his heart and love his neighbor as he loves himself "!**

Many years later, another man tried to deny self by his own will. He said," **The things that I don't want to do are the things that I find myself doing and the things that I want to do I do not.**" He found a law working in his body that is just as powerful as the laws of physics. This law fought against his will to do good. This man also echoed the Cries out of the past. He cried. **"Oh wretched man that I am. Who will deliver me from the body of this death?** This man was the apostle Paul. <u>However, he had found the answer.</u> He went on to say: **" I thank God through Jesus Christ our Lord. So then with the mind I serve the law of God; but with the flesh, the law of sin. For there is now no condemnation (death) to them which are in Christ Jesus. For the law of the Spirit of life in Christ Jesus has made me free from the law of sin and death."** Those men like Buddha and Solomon who lived under the law of sin and death ended up crying out, "we need a savior". No wonder Jesus said that Nineva and Sodom would stand up in judgment against this generation. They were like Buddha and Solomon, because they did not have the light and Life of Christ in their day. Until Jesus came and opened the door to life there was only **NOTHING..... VANITY....... DEATH!!!**

Jesus said that we must be born again to enter the kingdom of God where there is **eternal** life. Real life! John 3:36 says: **He that believes on the son has life. He that believes not on the son**

shall not see life, but the wrath of God abides upon him. By what we know about entropy and the eventual fate of the universe, this is very true. Reincarnation is merely a pipe dream. Jeremiah once said, "cursed is the man who trusts in the arm of flesh". No man has ever invented a perpetual motion machine, no man ever will. This writer once taught a class on the theory of electronics. He asked the class if a generator hooked up to a battery that was feeding it's electrical energy back into the battery, would run forever. They said that it would not, but no one could say why. The reason is entropy. Heat is lost in the wires, eddy currents in the generator, and chemical reactions in the battery. This heat is in the form of entropy and it cannot be recovered. Christianity is based on a **new birth** in Jesus Christ. Most all other religions are based upon salvation through flesh. The universe that our God created is no perpetual motion machine. It is a system of death. Perhaps before Adam sinned it was perpetrating itself, but now it is dying. Jesus Christ is the only door out of this system of death. Only those who believe on him and keep his commandments will ever escape from this death. His resurrection is the **only** proof of real life given to a world of death.

Thus, we see, that the sun is a type of Jesus the son of God. It sets in the evening in the color red. This symbolizes his shed blood at calvary and his death on the cross. It rises in glory in the morning to symbolize his resurrection to glory. But there is another light in the heavens who rules over the darkness. This is the moon. The light of the moon is a distorted reflected light of the sun. It's subdued half light makes things appear unreal. The light of the moon is a light for the predators of darkness that prey on the flesh of the weak and unsuspecting. Have you noticed how the moon tries to deceive us by imitating the sun and its seasons? It will come up first in one spot and then in another as it seems to move from the southern to the northern sky each month just as the sun does in its seasons. Can you think of an evil being who deceives men into thinking he is something that he is not? Of course! The moon is a natural type of Lucifer. His false prophets are the carnivorous beasts of darkness that prey on the weak. These evil people are his dupes. They deceive men by saying evil is good and good is evil. Have you noticed how they tell the homosexual that his perverted sexual preference is a biological trait instead of sin? Most of us have found that sex is a **choice.** This may not be true of a beast but mankind **chooses** to have sex or he chooses not to have sex. These false prophets of satan can call homosexuality a biological trait and every one believes it. Yet, if some man is caught in the act of adultery, that is called a sin. The judge calls it a sin too, as he hands out the penality of alimony payments. God has never called blue eyes or brown hair a sin because these are biological traits. One cannot help what color his eyes or hair is, but God has said that fornication, adultery, and homosexuality is a sin because it is a choice, not a biological trait. This is also true of women who turn from the natural use of their bodies to unnatural sex. We can choose to have sex or we can choose not to have sex. The only thing that anyone can truthfully say to the homosexual is; **"Repent of your sin and turn to Jesus Christ"**.

These false prophets say that God is a God of love not a God of judgment. We know that justice itself comes of love. Don't you want justice served on those who harm the ones you love? God loves his children, too. That's what our judicial system is all about; the jury judges, then comes the sentence, and justice is served. Justice itself is a product of love but judgment must come first before justice can be served. Don't believe these evil prophets of satan. Justice without judgment is impossible. They will cry pitifully all night for the life of a mad dog killer yet they will kill thousands of babies without ever considering the lives of these innocent children. Those women who approve of abortion say they have the right to do what they want to with *their* bodies, but they never harm **their** body. It's the body inside them that they harm. Some people say that abortion will allow the human race to clean up its genetic pool by allowing the "beautiful" people to have the children. How many Einstein's have we had out of the Albert Einstein Family?

How many Thomas Edison"s, Erico Fermi's,? Did all the great inventors and scientists of the past come from the rich and affluent? Indeed not! It's God who raises up and God who abases. These people would improve the genetic pool will eventually end up with a group of sterile, mentally retarded, sickly, zombies. These evil prophets of hell, are liars. Repent and be baptized in the wonderful name of Jesus Christ for the remission of your sins. Then stay in the race to the end and obtain a crown of eternal life. God has said that these stooges of satan will bring about their own damnation.

We have seen that the sun and the moon are types and shadows of Jesus Christ and the devil, respectively. However there is another type that must be considered. The scriptures say that a godly man is like a **tree** that is planted by the river of water that brings fourth his fruit in his season.(1st psalm) Each generation since Christ has had great saints who have led their generation into more light. The apostle Paul, Clement, Barnabus, John Huss, John Wickliff, Martin Luther, John Bunyan, John Fox and many untold others. Some of these Great men lost their lives for the name of Jesus, others spent many years in cold, dank prisons. But, because of their sacrifices much fruit was brought fourth and the church received back the light that was lost in the dark ages. They rendered great service for the salvation of mankind. Have you stopped to think about a tree and it's usefulness? Just like a saint of God it's whole life has been one of reaching heavenward for light. If you go down into a deep hollow in the mountains you will see the great, tall trees. They were born as little saplings in the dim light and shadows of the deep hollow. Now, they are great tall giants; their tops are bathed in the light of the sun. How useful these trees are to mankind as they sink their roots deep into the ground thus holding the soil in place to prevent destructive erosion. The man who sinks his roots deeply in the word of God will hold the soil of God's love firmly in place. A tree's green leaves remove the carbon dioxide from the atmosphere and breathe out pure oxygen. Without this purification process by the trees, man would soon suffocate in his own wastes. Without good, Christian, men who stand against the violence and pornography in our society, we too, would soon die in these evil wastes. In the fall these same leaves put on the most colorful show on earth. People drive hundreds of miles to see the beautiful colors of the Appalachian hills. Later these colorful leaves turn brown and fall from the trees. On the ground they form a mat of insulation to protect the young seedlings from the cold. They form small cups that hold rain water. There are so many of these small cups that it keeps millions of gallons of water from running off into the streams during winter rains. This helps in flood control. These leaves also rot and form the soft spongy loam that holds moisture and prepares the ground for the seedlings to make future forests. It's branches offer protection for birds to build their nests in. These birds eat hoards of insects that would soon devastate our crops. These same branches that offer the birds protection also bear fruits and nuts to feed both animals and man. Even in death the tree is useful. Lumber and firewood are used by man for shelter and warmth. Hollow logs make shelter for animals in the hard winter months. These logs eventually rot to enrich the soil. During its whole existence and even in death the tree is useful for both man and beast. This is typical of the great men like Paul, Peter and John who were the instruments God used to lay the foundation of the church and even in death their words are received with great joy by God's people of today. Indeed, they were the trees planted by the river of waters who brought fourth their fruit in their season.

Now you know that the tree is a perfect example of the godly man. In his true uncompromising way he bought the Bill of Rights with many years of imprisonment, some times with his blood. The James's, the Stephens', the Paul's, the Peter's, the John Husses, and the many others have suffered and died in order to speak God's word to their people in their generation. Some will say that our Bill of Rights was a product of the renaissance but the renaissance started in Italy, if this were true, why didn't Italy have the first democracy? A democracy in Italy was unheard of until after the 2nd world war. When our forefather's framed the Bill of Rights. I do not believe that they had the renaissance in mind. It was the martyrs who lost their lives for the name of Jesus, throughout Europe during the reformation, that they were thinking about. These brave men and many like them have prepared the way for each new generation of godly men as they reach high and tall toward heaven to receive the light of Jesus Christ and the wisdom of God hidden in Him.

Just as most trees bear good fruit so does the godly man. His fruit is manifested in the life that he lives. The fruit of his society is shown by the political system that he brings about. Let me ask you a question; how many non-Christian countries have a bill of rights guaranteeing human freedoms? By the same token; how many non-Christian countries even had a democratic form of government before the 2nd world war? Jesus Christ said, "By their fruits you shall know them." Their political systems are a direct reflection of the way they believe. Even in Christian countries whose religious churches are dictatorships who claim that the church and the men who run it are infallible have had one political dictatorship after the other for centuries. Not many countries even now allow freedom of worship. In our own country, freedom of worship is in danger because many well-meaning people are being duped by satan's false prophets by the fear of loosing our separation of church and state that is guaranteed by the constitution.. Even though these great godly men did win our personal freedoms, any country that falls away from God into disobedience, soon begins to loose these hard won freedoms. It has become evident to us that satan does not quit in his effort to stop the purposes of God. In this country we have the right to speak, write, assemble and worship as we please. The devil knows that he can no longer persecute and kill the saints who live in countries who have these basic freedoms. **Instead he uses these freedoms to bring about moral and religious anarchy**. He has "liberated" women from their natural state of motherhood and making a home for their children into trades that were once the sole domain of men. We find many women in the construction trades, engineering, the military and other masculine jobs. Because of this, the family unit is being destroyed. Where once our children received their spiritual guidance from their Christian mothers, they now get "brainwashed" by the medium of television and ungodly, secular schools. Wherever satan can, he holds people in the bondage of evil dictatorships. But when they win their freedom he then begins to dupe them into a state of anarchy, so that "law and order" will take them back into the slavery of dictatorship. This is happening now in our own precious country. How much longer will we have our freedom while evil beasts roam our streets and highways murdering people at will?

So called "freedom societies" have liberated us into the freedom of pornography yet they have guarded our school systems from anything that resembles the freedom to worship. Our children can sit in our living rooms and watch all kinds of vile sex and violence on the television yet they cannot even mention the name Jesus or say a prayer in their school. One school in

Kentucky faced litigation in the court system because the school board persisted in leaving the ten commandments on the school room walls. However, we find that most societies in the world have incorporated the ten commandments into their laws. Even the non-Christian nations have laws against murder, adultery, larceny and most of the other commandments. These laws are the cohesive force that holds a cooperating society together and keeps them from becoming beasts. The ten commandments were given to flesh to keep and preserve society until a savior came. Yet these evil, satan inspired, people persist upon taking us further and further into anarchy. **Freedom is good and it should be vigorously supported, but only if it is for the common good.** Freedom that is for personal gratification, tears down the very moral fabric of our nation. The good people of our nation should stop this moral degradation before it is too late. The false prophets of satan call vile pornographic smut "art" under the immoral guise of realism. This author was taught in his college humanities classes that art was supposed to have a redeeming effect upon society. Charles Dickens opened the eyes of the British people, and the world, to the plight of the poor. His literature was the beginning of a movement for a Middle class. It was a movement away from evil dictatorships and kings who ruled by "divine right". Other artists have, in their way, tried to reform society all down through the years. Some art has been a good influence and some not so good, but our modern day, "art" is nothing but vile smut. It is taking society into the gutter. And some of this so called, art, is supported by the tax payers themselves. What a crime against the very fabric of society!

We have seen how the devil's false prophets have brought about anarchy in our morals by saying good is evil and evil is good but what about our churches? The woman (church of Jesus Christ) of Revelations chapter 12, escaped on the wings of this country's symbolic eagle into the **wilderness** of the American colonies, there she obtained her religious freedom. The devil has been busily subverting this hard won freedom by casting a great flood of humanism, evolution, and Freudian theories of the beast out of his mouth in order to sweep the woman away. Read the 12th chapter of revelations. Psychology and psychoanalysis has invaded our churches. Pastors no longer use the word of God to heal trouble minds, they use psychology instead. Evil pharisee religions push for dictatorships where they can rule and dominate the people's worship. Our world is moving toward this religious system to prepare the way for the anti-Christ. These evil pharisees are behind the great falling away. We have watched our belief in Jesus Christ become eroded by this great flood. Along with this has come a dose of eastern religious, mysticism. It has become the modern religion of the beast. We find that the only thing these churches are committed to is **self**. That's why we hear so much about self-esteem. Freud, known by many as the father of psychoanalysis, taught that whenever we frustrate the natural instincts of procreation,(sex) and survival, (selfishness) we become mentally ill. We know from our observation of animals that these two things are the primary concern of a beast. Therefore, when we teach these things in our schools we are teaching our children to worship the beast. This may sound pretty strong to you but these precepts are based upon the **theory** of evolution. Now, theory is not **fact**, it is theory, therefore, it is more a religion than it is a science. "Philology recapitulates ontogeny" is one of the many "proofs" of evolution. This is the theory that the human embryo recapitulates the various stages of evolution during its development. In this book we have seen that the very nature of the universe points to Jesus Christ. Is it any less "scientific" to believe in the types and shadows of him found in the creation? Suppose a teacher in a biology class in one of our secular schools pointed out to his class that in order for a man to live he must eat. But when he eats, something must die. If he eats meat, an animal dies. If he eats vegetables, a plant must die. He then tells his class, in order for man to live something must die. He says; this is Nature's way of pointing to the sacrifice of Jesus Christ. He Died that we might have eternal life. The A.C.L.U. would probably protest, litigation would ensue, and the teacher would be fired. But what if another man said, this is nothing but "survival of the fittest". The strong eat the weak. He would go on teaching into his old age, wouldn't he? But who is right, the evolutionist or the Christian? Both of these ideas seem to have credence. Evolution is not a science it is a religion. It is an accumulation of things that point to evolution, but my bible is also an accumulation of things that point to Jesus Christ, however, I must accept them by faith. Any belief that cannot be proven in the laboratory is a religion not a science. Evolution is the gospel of the beast. Perhaps soon, other great **"trees"** will have to die because they stand against this deceiving lie. This writer believes that the precepts of evolution was purposely put into the creation by God so that these evil dupes of satan might make up the great lie of the beast and become dammed. 2nd Thessalonians 2: 11

It's all right to teach Yoga (eastern mysticism) as a method of relaxation in our schools, yet we strictly forbid a school prayer. Why is it that our federal lawmakers and supreme court judges with their degrees in law and political science make such a difference between the occult methods of eastern religions, that is taught in our state supported schools, and a simple Christian prayer? Do you, as an ordinary citizen, think that you can tell a real danger to the principle of separation of church and state? Suppose your child came home from school and announced that he was told that unless he became a Methodist he would be kicked out of school. Would you say this violates the principle of separation of church and state? Of course you would! Suppose tonight's paper said that any government employee seen going to church would be fired; that's another real danger isn't it? Now, Let's say you came home from work tonight and a big headline in the paper said, **Chaplain opened Congress with prayer! Atheist house democrat protests to the A.C.L.U., Threats of court action ensue.** Of course this probably won't happen because our constitution states that each session of congress will be opened by prayer. This fact **strongly** suggests that the ones who framed our constitution were not against prayer in a federally supported institution doesn't it? They figured that the people themselves would have the wisdom to tell a **real danger** to the separation of church and state. Perhaps our educational system is worse off than we thought. What has happened to this generation? Have they lost the wisdom to tell a real danger to church-state separation or are we subtly being deprived of our freedom of worship by a very subtle group of globalists? Some people give the excuse that little things lead to bigger things, pretty soon the government will be telling us how to worship. I repeat, have we lost our wisdom? We are so highly sensitive to church-state separation yet completely insensitive to Pornographic smut. Do you know that our constitution strictly forbids the sending of pornography in the united states mail? This is strong **inference** against pornography isn't it? How many cases brought before the supreme court have been decided by inference? How many federally regulated systems of communication did we have back then? Only the mail, right? Now we have federally regulated television, radio, telephones, computers and many other methods of communication. Do you think that this is strong **inference** that our forefathers would have forbidden pornography in them as well? **Of course you would.** But our federal judges don't. Our federal lawmakers don't. And the A.C.L.U., who pretends to guard so jealously our constitutional rights doesn't either. Why are our Supreme Court judges and other federal judges so very sensitive to religious overtones in our public schools yet so insensitive to the pornography that floods our federally regulated communication systems. Many laws made by congress have been declared unconstitutional by the supreme court on the grounds of INFERENCE. How strong does the inference against pornography have to be before our judges stop this flood of pornography that inundates our federal communications systems. Industrial poisons are threatening our water supply. **A polluted river can be cleaned up but only God can clean up a polluted mind.**

Recently our congress had a proposal before it to supply federal money to private schools. Suppose that these private schools became dependant upon this federal money. Then some bureaucrat could threaten withdrawal of federal money if they didn't teach a federally approved religion. Suppose he announced to a fundamental church organization that they must start teaching evolution against their will or loose their federal money. This would be a real danger to separation of church and state wouldn't it?

We have expelled prayers from our schools on the basis that they are federally funded. Yet many of us in this country have FEDERALLY FUNDED home mortgages. Does this mean that if some atheist walks into our living room (heaven forbid) and sees a bible, and protests to the ACLU, that we will have to get rid of our bibles? This is the same insane reasoning that is operating in our judicial systems right now. How much longer must we sit back and watch our religious freedom being taken away from us under the guise of separation of church and state? Pretty soon, there won't be any religious freedom. It sounds like a sinister plot to wipe out true Christianity and replace it with a bland one world religion doesn't it?

We see evidence of the subtle work of the devil as he has undermined the right to worship as we please. This right was won by the great godly man (**trees**) of past generations who brought fourth his knowledge of Christ in his season. The devil has brought about anarchy in our moral and political values while at the same time undermining our right to worship almighty God. If a Christian decides to educate his children at home or in a private church school according to Christian values he is harassed and oppressed and even thrown in jail. This actually happened in the case of Nebraska vs. Reverend Silven. He was put in jail without a warrant on one occasion, taken right out of the pulpit during a church service on another occasion, and deprived of other civil rights as well. During this time, no one came to his aid. I have heard of freedom societies defending the civil rights of a communist, a fascist , a pornographer and even a child molester, but never has this writer heard of them defending the rights of a Christian to worship as he pleases. We need Godly "trees" to stand up for truth and our Christian rights in this generation as never before. Perhaps it's too late.

In this chapter we have seen that the sun who is the life giver of the earth is nothing but a type and shadow of Jesus Christ the eternal life giver of God. Just as the sun brings about the weather to cause sustaining rains that bring about plant life to furnish food for man and animals, Jesus Christ causes the Holy Ghost rain to fall on Christian believers to sustain them against the wiles of the devil. We also saw that the moon is a type of the devil with his deceiving, unnatural, reflected light that causes men to believe a lie that they might be dammed. The great trees that bring fourth their fruit in their seasons are a type of the Godly man who has stood in the breach for all peoples down through the years. They brought fourth the freedom we now enjoy by their suffering. As we move away from God we are beginning to throw away these freedoms. We are promoting freedom for self gratification not freedom for the common good! May God forgive us and reinstate us into his wonderful love. AMEN.

CHAPTER FIVE

VIRGIN BIRTH....A NATURAL EVENT

As you have probably guessed by now, this writer has endeavored to show how the purpose and plan of salvation was made known by the things that are made. Not only in this world but by the very nature of the universe itself. There is one other thing that I would like to show you that was a natural event, however, the bible SEEMS to show that it is in the supernatural realm. But when you take the bible as a whole from Genesis to Revelations and consider the prophesies and covenants we come to a different conclusion. When we read the bible we must take it all not just one isolated incident. If it ALL doesn't say the same thing then something is wrong. When you read this remember that the church which the apostles founded was not the same church after Constantine and his sun worshippers took it over. There is much evidence that things were added to the old manuscripts during this time. The event I am speaking about is the virgin birth of Jesus Christ. Jesus Christ was the word of God. He didn't say, I speak to you the truth, he said, I am the truth. Please hear me out and you will find that there are many scriptures in the bible that bear up the fact that the birth of Jesus Christ was a natural birth. There is strong evidence for this by the absence of any mention of the virgin birth by any of the apostles in their letters to the churches.. Neither John, James, **(THE LORD'S NATURAL BROTHER),** Peter, or Paul ever mentioned the virgin birth in any of their epistles after the resurrection. Neither did the church fathers, Polycarp, Clement, or Barnabus ever mention it. The only mention of a virgin birth was by Ignacious, and his writings were not found until four centuries after Jesus's death. Doesn't this seem quite strange to you? A virgin birth is unknown in nature yet no one in the early church mentioned it. Paganism has had a **mother-god-queen** figure in it since antiquity. She was called the **"queen of heaven"** in the book of Jeremiah. The apostle Paul encountered her as Diana. He was beaten with many stripes and thrown into prison because of her. From historical accounts the virgin birth did not appear until several centuries after the resurrection of Jesus Christ. It was concurrent with the entrance of Gnosticism into the church. The Virgin Mary has replaced the "queen of heaven". Contrary to the word of God she is worshipped in the Catholic Church. Just remember that the word of God cannot contradict itself. Jesus Christ was WORD. **God in the flesh**. All that was spoken by the prophets was given by the Holy Ghost. The Word is the very power of God. It had to come to pass **EXACTLY** as it was spoken. I will now endeavor to show you that Jesus came into this world through the **SEED** of men that God made **SEED** covenants with. Jesus said, in the spirit there is neither male nor female. Therefore, seed must come from flesh not spirit. Remember, Matthew 4:28: The earth bringeth fourth fruit of itself. The Holy Seed was brought fourth by the power inherent in the word of God. Seed covenants were made by God with Eve, Abraham, Phinehas, David and with the seed of a people called the Rechabites. A covenant with God cannot Fail, And I will show that these covenants did not fail.

We have all heard that Jesus Christ is word, but do we really know the awesomeness of this fact? John the apostle said , **In the beginning was the word and the word was with God and the word WAS God.** In the beginning of what? God always was, there was no beginning of God. It was the beginning of the creation. Jesus was co-existent with God even before the creation. Remember that God is omnipotent (all powerful), for this reason his word **MUST** be truth. If God says something is going to happen, **IT HAPPENS**. The prophets spoke by the power of the Holy Ghost. What they have predicted has either happened or **it will happen** at its appointed time. All the predictions concerning Jesus Christ came to pass and I will show you that they came to pass during his life on earth. The things that **Jesus predicted** will come to pass at the appointed time as well. Because he is word, this is a sure thing! There's not demon in hell nor any demon possessed man on earth that can stop it. As an example, God spoke to Daniel 1500 years ago and said, *at the time of the end knowledge shall increase and men shall run to and fro.* Has this happened? Look about you, people travel all over the globe in a matter of hours, knowledge increases so fast that it is impossible to keep up even in your own field of work. This is only one example. Israel was made a nation in 1948 against unbelievable odds, this was predicted many thousands of years ago. There are hundreds of other examples of the word coming to pass not only in this day but in times past as well. He has spoken, in the book of Revelations, that at the time of the end there will be a world government ruled by the anti-Christ. We see that it is in the process of formation right now. Jesus himself said, **every dot and every tittle will be fulfilled**. These prophesies and those that follow are an attempt to show you that whatever God has spoken has come to pass or must come to pass in the future. **They cannot fail.**

God revealed to John in the book of Revelations that a beast would rise up out of the sea.(Revelations chap 13:1) In the bible the word sea symbolizes peoples. Communism was born by the wrath of the people in the French revolution. The people were tired of being mistreated by evil selfish kings and a corrupt church that cared nothing for their plight. They rose up and overthrew the monarchy and the church. The people instituted the first communist government. **<u>Communism has called itself a dictatorship of the peoples.</u>** In the bible, the word "sea" used in symbolism means peoples. This communist beast rose up out of the wrath of the people. John wrote in Revelations that this beast would hate the whore and would persecute her and burn her with fire. The word whore is used here as a symbol of the church that compromised the words of her Lord in order to obtain money and power. A whore sells her body. This church was a corrupt representation of the body of Christ. Communism has been atheistic from the start and has persecuted the state established churches wherever it has taken over. This communistic Russian beast appears to have fallen, but it still has many things yet to fulfill. However, it is no longer dominating the world scene. It came to fulfill the word of God and bring wrath upon the whore church that spawned it. At this time Russia and the other communist nations are no longer threatening world war.

John also spoke of another beast. The next beast John speaks of would rise up out of the <u>earth.</u>(Rev. chapt. 13:11) The first beast rose out of the sea (people) but this beast, will rise out of the political and religious systems of the earth. The evil Babylonian system has been **in the earth** since the time of Nimrod. This religious Babylonian system has opposed God and his word (plan) all down through the centuries. We see our beloved nation turning away from god into mystery religions, with it's astrology, mediums, fortune tellers and psychics. These psychics advertise very boldly on television. However, God says they are an abomination to him. We no longer believe in Jesus Christ as our salvation, instead we believe in reincarnation and the humanist doctrine of evolution.

95

These things should make us tremble because of the coming wrath of God. God spoke to Paul and said that at the time of the end there would be a great falling away. This statement is not made about sinners, men of sin have nowhere to fall from. This statement is aimed at the church. Since the advent of evolution, psychology, and scientific knowledge, many churches don't even pretend to hold fast to God's word. Instead they preach a demon inspired ecumenical gospel of one world religion. They no longer say that Jesus Christ is the only way to the Father (God Almighty). They no longer preach against sin, instead they say good is evil and evil is good. Sodomites are not only allowed in these churches, but they are allowed in the pulpit itself. What a stench this must be in the nostrils of God! What God spoke to Paul **had to happen** and it **is** happening. Again, we see that god's word is true because there is a great falling away going on right now that far surpasses any thing that has taken place in previous times. All these things that are happening in this present day are direct evidence that God's word **cannot fail**.

I have shown you this because God is omnipotent and the seed covenants that he made with flesh concerning his Son Jesus had to come about exactly as the prophesies said they would. **God's word is truth. It is always truth!** It is impossible for the things that an all powerful God says, to fail. Because of this, his word is the only thing in the universe that is truth. By definition, truth

96

must be absolute. It cannot change. It must stand for all eternity; unchangeable forever. If it doesn't fit this criteria it is not truth! When we look into scientific truth alone, technology and the laws of physics will, someday in time, **be folded up as a garment,** all will be ruled by absolute zero....death! God has said this and science backs it up by the laws of thermodynamics. The 2nd law of thermodynamics says that all systems that do work are loosing energy that cannot be recovered. This energy is called entropy. The entropy of the universe is increasing. This means that the universe is dying. God is eternal and only his truth is eternal. Jesus said, **"I am the truth".** God has spoken through the mouth of his prophets and said the heavens and the earth shall wax old even as a garment and they will pass away, but he says, my word will never pass away. Therefore nothing in the universe is real truth except his eternal **WORD.** All else is passing away. Whatever God has spoken through the mouth of his prophets has either come to pass or it **WILL** come to pass at it's appointed time.

Who or what is God's word? Remember John's gospel (chapter 1) says: *in the beginning was the word and the word was with God and the word was god. All things were made by him and for him and without him nothing was make that was made.* Therefore Jesus is the WORD. Those who are in his eternal body are word as well, **All else is death!** Psalm 1:4 says: *The ungodly are not so but they are like the chaff the wind drives* *away.* Many believe this is figurative but it is a literal fact. We are born in a system of death and we will stay in this system unless we believe and enter into the WORD, Jesus Christ. He said, **let the dead bury their dead, come, follow me.** If you do not believe that Jesus was with God from the beginning read the 8th Proverb starting with verse 22. The 8th Proverb shows us that even before the creation Jesus Christ, the wisdom and power, was with God and, because he was word, he was God. This man Jesus who was God manifested in the flesh said, **"I am the way, I am the truth and I am the life; no one comes to the Father but by me.** He didn't just say I'll show you the way, many false or ignorant teachers have endeavored to show the way. His Statement was much stronger than this; he said, in effect, there is no other way! He also said**, I am the truth.** Buddha once said that he was a seeker of truth, but he never once said; I am the truth. What man on this earth would say, I am the truth? He would have to be a man with delusions of grandeur. However, Jesus was in full command of his intellect and emotions. Jesus so confounded the intellectuals (scribes) of his day that they dared not ask him any more questions because he was making them look like fools. He also said , **I am the life!** He didn't say I'll show you how to live. Jesus Christ knew the laws of thermodynamics, he ought to, he made them. He knew that we are all flesh, part of a physical system of death. He knew that we must be born again in order to enter **real life.** Only in God, is there real life; life so wonderful and glorious that our minds are unable to comprehend it. Jesus said I am the truth, and so he is, because God's word cannot be anything but truth! Why? Because of his omnipotence. Therefore, what the prophets spoke about Jesus had to be truth. Please remember this. ALL that was spoken concerning Jesus had to come to pass. **THERE ARE NO DISCREPANCIES.**

The old testament is full of prophecies concerning Jesus, let's look at some of them. (Deuteronomy 18: 18) God is speaking to Moses concerning the coming of "that prophet" (Jesus): **I will raise them up a prophet from among their brethren like into thee: and I will put <u>my words</u> into his mouth. And he shall speak unto them all that I <u>command him to speak</u> and it shall come to pass that whosoever will not harken unto all that he shall speak in my name I will require it of him.** This indicates that Jesus would be a flesh man just like Moses. That he would be an Israelite and of Moses' brethren. Many in this day refuse to accept the word of God, Jesus, as the savior of the world, instead they declare that he was only one of the masters, like Buddha, Krishna, or Vishnu. They say he didn't bring mankind salvation but he only came as a great teacher to show us how to live. But Jesus actually declared to the pharisees that he was "that prophet" sent by Moses' God. Look at the 12th chapter of the gospel of John starting with verse 47: It says, **If any man hears my words, and believes not,**

I judge him not for I came not to judge the world but to save it. He that rejects me and receives not my words has one that judges him. The words that I have spoken, <u>these shall judge him in the last day.</u> Do you see how this confirms what God said to Moses," that man who does not harken unto all that he shall speak in my name I will require it of him." God will require it of all men in the judgment on the last day. Jesus, however went further in vs. 49, he says: **For I have not spoken of myself, but the Father which sent me (Moses' God) gave me a <u>commandment</u> what I should say, and what I should speak.** Jesus is confirming the very words that God spoke to Moses centuries before in the 18th chapter of Deuternomy. He spoke those things God **<u>commanded</u>** him to speak. These words were spoken to the very ones who claimed to be staunch supporters of the laws of Moses, and they wouldn't hear him. It was hidden to them because Isaiah had said, **"Hearing they will not understand."** This certainly supports the fact that Jesus is indeed the son of Moses' God, not some second rate master who went to join his karma. **<u>He is the master.</u>**

By now I hope that you see that Jesus was word and that all things said about him by the prophets was truth. **<u>This had</u>** to be because he said, "I am the truth". For this reason, let us look at more prophesies that were made concerning Jesus. These are confirmed in the new testament gospels:

Psalm **22:1 My God, my God, Why have you forsaken me. Mark 15:34**

Psalm 22:16,17,18: For dogs have compassed me. The assembly of the wicked have enclosed me: They part my garments among them and cast lots upon my vesture. Mark 15:24

Psalm 16:10 For thou will not leave my soul in hell neither will thou suffer thine Holy One to see corruption. (Acts 2:27) Jesus arose from the grave. He arose in the flesh. He did not manifest himself as a spirit. He ate and drank and still bore the wounds of the cross in his body, not only this, he ascended to the father in the flesh. He did not fly out somewhere to join some nebulous entity called karma. He even now sits on the right hand of the Father. Stephen confirmed this while being stoned.

Psalm 18:43,44 Thou hast delivered me from the striving of the people and thou hast made me a head of the heathen: a people whom I have not known shall serve me. As soon as they hear of me they shall obey me. We all know how fast the gentile church grew. In connection with this, see John 12:20-25. **And there were certain Greeks among them who came to worship at the feast. The same came therefore to Philip of Galilee and wanted to see Jesus. Philip then told Andrew and Andrew and Philip told Jesus.** This is what Jesus answered, **The hour is come that the son of man should be glorified. Very sincerely**

I say unto you except a corn of wheat fall to the ground and die it abides alone, but if it dies it brings fourth much fruit. Why did Jesus answer them this way? These Greeks knew that the pharisee's were plotting to kill him. The Greeks wanted him to go with them to Greece to be their teacher. There he would be safe. But, Jesus knew that this is what they wanted to see him about and this is why he answered them the way he did. He knew that he had to fulfill all that was written of him in the law, he knew that he had to die that the salvation of the world might be brought fourth as the fruit of his death. Later, the Holy Spirit sent Paul to these Greeks, and at that time, the gentile church did grow very fast, as soon as they heard the word, Psalm 18 came to pass.

Psalm 41:9 **Yea mine own familiar friend in whom I trusted, which did eat of my bread, hath lifted up his heel against me.** As soon as Judas ate the sop of bread he went out immediately and betrayed him. Isaiah 11:10-11 **And in that day there shall be a root of Jesse, which shall stand for a ensign (symbol of authority) of the people; to it shall the gentiles seek and his rest shall be glorious. And it shall come to pass in that day that the Lord shall set his hand the SECOND time to recover the remnant of his people which shall be left, from Assyria, Egypt, Pathros, Cush, Elam Shinon, Hamath and from the islands of the seas.** Some Christian cults in this country claim that the Jews who are now in Israel are not really Jews. They say that this prophesy is talking about the dispersion into Babylon. We know that the Jews were taken captive into Babylon some time around the 6th or 7th century BC. But they returned to Jerusalem 70 years later to build the temple for Jesus. The above prophesy was speaking of a much greater dispersion. Babylon is almost due east of Juda, but the above prophesy speaks of a dispersion much greater than this. This prophesy has come to pass in our own time. Some of these cults say they are Jews. If you still have doubts about the Jews of this day read Romans chapter 11; 17-24. **Cursed are they that say they are Jews and they are not.** It would be best if you read the whole chapter. It tells of the grafting in to the tree the wild branches. (Gentiles) In 1948 Israel became a reality as it became a recognized, independent, nation by most of the world. Six million Jews had lost their lives in Hitler's ovens. It looked as though the Jews were completely decimated and could never recover. Out of this utter defeat millions of Jews made a great exodus to Palestine, and God's word came to pass. Just when it looked like the devil had won God was victorious, Praise his great name! The nation of Israel became a reality against great odds. If these people who are now in Israel are not really Jews it seems very strange that the Devil is going to such great measures to destroy them from the face of the earth.

Psalm 16:**10 For thou will not leave my soul in hell neither will thou suffer thine Holy One to see Corruption.** Peter used this psalm in his first sermon to prove the sonship of Jesus to the church. (Acts 2:27) Jesus arose from the grave. He arose in the flesh. He did not manifest himself as a spirit. He ate and drank and still bore the wounds of the cross in his body, not only this, he ascended to the father in the flesh. These are just a few of the many prophesies found in the bible concerning Jesus that have come to pass; we cannot cover them all. However, I hope you are beginning to see why Jesus was able to say, I am the truth.

There is another form of prophesy found in the bible which is defined as the combination of a promise and a binding agreement between two parties. God calls it a **covenant.** These covenants were **seed** covenants and they established who Jesus's ancestors would be. The covenants discussed here will be between God and three men, Abraham, Phinehas and David. Also, the first woman, Eve, and a tribe called the Rechabites were included in these "seed" covenants.. Note that if God makes a covenant it cannot be broken. The first one is found in Genesis 3:15 God is speaking to the devil: **And I will put enmity between thy seed and her seed It shall bruise thy head and thou shalt bruise his heel.** We all surely know who bruised the devil's head when He rose from the dead.. The second covenant is found in Genesis 17: 7 with a man named Abraham: **And I will establish my covenant between**

me and thee and thy seed after thee in their generations for an everlasting covenant to be a god to thee and to thy seed after thee. God said that all the families of the earth would be blessed by the seed of Abraham. Jesus Christ came from the seed of Abraham and we surely have been blessed through Jesus. Again in Genesis 12:3; **Sarah thy wife shall indeed bear you a son. I will establish my covenant with him for an everlasting covenant, and his seed after him.** Notice that God established this covenant with Abraham's son and his seed after him. Abraham's son was named Isaac. Isaac is a prototype of Jesus and the body of Christ is the seed of Jesus.. This promise is what we will call a **seed covenant**. This is the first of other seed covenants that will be discussed in this chapter.

The next seed covenant is found in Numbers 25:11; It would be best for you to read the whole chapter for your self. This account tells of a covenant made with a Levite priest named Phinehas. Please do not confuse this Phinehas with the Phinehas God slew along with his brother for their disobedience. (Samuel 1:3; 2:34) The evil Phinehas was the son of Eli. The covenant with God was made with Phinehas, the son of Elezear. He was a grandson of Aaron, the brother of Moses. In Deuternomy 18: 18; God spoke to Moses and said, "I will raise up for them a prophet of their brethren like unto thee. They were all descendants of Levi, therefore, they were the brethren of Moses. This covenant

105

was mentioned in another chapter of this book but in case you haven't read it I will relate the important points to you. God had warned his people to stay away from idol worshippers. God had the same motive that modern men have when they move from large cities and good paying jobs to small rural communities. These men want to get their children out of the drug infested, high crime environments of large city schools. God's motive was similar, he did not want his children to become tainted with evil idol worship and its immoral, sodomite ways. But just as children disobey parents, so the children of Israel disobeyed God. They went ahead and started mixing with these Middianite men and women. Because of their disobedience, God sent a deadly plague on them. So, Moses called the congregation together to seek God about the matter. This was more or less like a church meeting. Suddenly right in the middle of the service an Israelite man and a Middianite woman paraded through the congregation to his tent to make love. This was a direct slap in the face of both God and Moses. Just then, a Levite priest named Phinehas seized a spear, ran into the tent and slew both the man and the woman. God later spoke to Moses saying; NUMBERS 25:10: **Phinehas the son of Eleazer, the son of Aaron the Priest has turned my wrath away from the children of Israel while he was zealous for my sake among them, that I consumed them not in my jealousy. Wherefore say, behold, I give to him my covenant of**

peace : and he shall have it, and his SEED after him, even the covenant of an everlasting priesthood because he was zealous for his God and made an atonement for the children of Israel. Here you have the 2nd **seed covenant.** It was with Phinehas the grandson of Aaron of the tribe of Levi. With this covenent, God established the everlasting Melchezedk priesthood with the tribe of Levi. Remember a covenant with almighty God cannot be broken. As a type and shadow of this promise to the seed of Levi, God established that only the Levites could carry the ark of the covenant. The ark was a type of Jesus Christ. David once broke this commandment and caused others than the Levites to transport the ark. In his anger God slew one of these men. (See 1st chronicles 13:10) In Hebrews 5:10, the apostle Paul says that Jesus is acting as our Melchizedek Priest, therefore, Jesus fulfilled the covenant as a direct descendant of Phinehas, because he is the one who established the everlasting Melchezedk priesthood. Therefore, the covenant between God and Phinehas was kept. A covenant with God must be kept. But how was it kept if his mother was a Jew not a Levite?

There is strong historical evidence that Mary's father was a Levite priest named Joachim. In the "Lost Books of the Bible", the book of Mary has a preface which makes the following statement: *"The book of Mary was attributed to the apostle Matthew. It was found in the works of Jerome who flourished about the **4th*** ___**century A.D.**___ *The more ancient copies differed from Jerome's, for from one of them, written by the learned Faustus, a native of Britain, who later became the bishop of Rieze. This man Faustus endeavored to prove that Christ was not the son of God until after his spiritual baptism and that he was not of the tribe of Juda because Mary herself was not of this tribe, but ___was of the tribe of Levi.___ The reason, he stated, was that her father was a Levitical priest by the name of Joachim."* I do not agree with this man in principle because I believe that Jesus Christ was the Son of the living God from the foundation of the World. But, I do believe his statement that Mary was a Levite and not a Jew. **I believe that the seed covenant with God and Phinehas was kept through Mary the mother of Jesus** who was a Levite not a Jew. Thus, the covenant for the everlasting priesthood that was promised to Phinehas was kept. Elizabeth was the mother of John the Baptist. She was of the Daughters of Aaron. Elizabeth was a cousin of Mary. This gives biblical evidence that Mary was a Levite.

We know that in the book of Isaiah ,chapt 7:14 it says: A virgin shall conceive and bear a son and his name shall be called Emanuel. Many sources question the use of the word virgin here. For example, the NIV Hebrew-English Bible states: *"The NIV Translators chose "The Virgin" when they translated the word but the choice of "young woman" would have been the better choice, linguistically, contextually, and theologically.* The NIV translators chose the word virgin because the King James translators had chosen the word virgin." This author is not a Hebrew scholar but I can show you the difference in the word for VIRGIN that was used in Isaiah 7:14; and the word for virgin used in many other places in the bible. The word used in Isaiah chapter 7 means "Young woman". The other word found in many places in the bible means **"never having been with a man".** The Hebrew word for YOUNG WOMAN looks like

THIS;

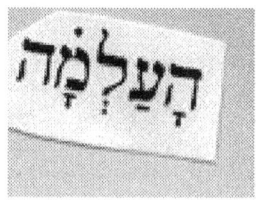

The word for a true VIRGIN found in other places in the bible looks like

THIS:

You can see for your self that there is a difference in the two words. Since the translators of the bible were more influenced by

the account of the virgin birth than they were the literal meaning of the word, I believe that the word, virgin was used instead of **YOUNG WOMAN**, which was the better word. My beliefs are based upon the accounts of **the seed covenants** found in the bible and are stated in this book. Indeed Mary was a young, pure woman who God chose to bear his only begotten son. She was the wife of Joseph, of the tribe of Juda. You will see later that Joseph's name was used in the genealogies given in Matthew and Luke, Mary's genealogy was not mentioned. This gives more scriptural evidence that Joseph was his flesh father Now in Jeremiah chapter 35:19 there is another seed covenant with a tribe in Juda known as the Rechabites. This is what the covenant says: **Therefore, thus says the Lord of Hosts, the God of Israel; Jonadab the son of Rechab shall never want a man to stand before me forever.** While searching for an account of how this covenant was fulfilled in Jesus Christ, I found this account in the encyclopedia Bratainica: "The Rechabites were a conservative, ascetic self denying Israelite sect that was named for Rechab the father of Jehonadab. Jehonadab was an ally of Jehu, a 9th century B.C. king of Israel, and a zealous antagonist against the worshippers of Baal, a Cannanite god. Though of obscure origin, the Rechabites were related to the Kenites, according to Chronicles 2:55, a tribe absorbed into Juda in the 10th century B.C.. The Rechabites were separatists who refused to participate in agricultural pursuits. drink wine, or engage in other activities associated with the Cannanites. Believing that the semi nomadic way of life was a religious obligation they herded their flocks over much of Israel and Juda. (According to the book of Jeremiah, they felt obligated to live this kind of life because their father Rechab had told them to do so.) They were fervent followers of Yahweh, the God of Israel. They are best known for their slaughter of the priests of Baal during the revolt of Jehu. (ll kings 10: 17-27) <u>According to later Jewish tradition, the Rechabites intermarried with the Levites, the priestly class.</u> This being true we can see the covenant with the Rechabites in Jeremiah was fulfilled due to their intermarriage with the Levite priesthood. This lends further credence to the fact the seed covenant with the Levites was fulfilled through Jesus' mother Mary who was of the daughters of Aaron.

There is one more SEED COVENANT that God made with David the king of Israel of the tribe of Juda. The account of this is found in 2nd Samuel; chapter 7. It is also found in 1st Chronicles chapter 17. At this time David's kingdom was stabilized. He had defeated his enemies and the kingdom of Israel was at peace. David began to think about building a great house to worship God. So he asked Nathan the prophet if it was all right to start building God a house. Nathan answered; **"Go ahead for god is with you."** However, that night God came to Nathan Saying, **"From the time that I delivered the people of Israel from Egypt, I**

have never said anything to them about building me a house. But instead I have gone from tent to tent and from house to house. God intended to have a house of flesh to dwell in, HIS VERY OWN PEOPLE. Jesus would be the CORNERSTONE of this house. Thus, he said to Nathan: **"Go tell David that he must not build me a house for I will build him a house. And it shall come to pass that when he has expired and gone to be with his fathers. I will raise up his seed after him. Which will be of his SONS and I will establish his throne forever. I shall be to him a Father and he shall be to me a Son**. (God made a **<u>seed covenant</u>** for an everlasting Kingdom.) We also see that this <u>seed covenant</u> would come to pass through his sons, **not his daughters.** If Mary was of the house of David she certainly was NOT of David's sons. This gives further scriptural evidence that Joseph was his flesh Father For God does not and cannot lie. Many people believe that this covenant was to Solomon for a great temple and that the kingdom was for a long succession of kings, nothing else. They think that the house God was talking about is nothing more than the great temple that Solomon built. Most of the kings who where of the house of David, were pagan idol worshipers. The genealogy that takes Jesus back to David found in Luke chapter 3; does not take the route through Solomon at all. See for yourself. It takes a route through some unknown men of David's blood back to Nathan, then, to David. Nathan was Solomon's full brother. Luke says that Jesus was a descendant of Nathan, not Solomon. While satan was busily seducing the kings of David's blood, in as effort to keep his word from coming to pass, God was establishing the house of David through the seed of Nathan, Solomon's brother. It doesn't matter what satan does in his effort to thwart God, he always looses.

You will also note that the bible translators stated the genealogy of Jesus this way: Jesus being about thirty years of age being the Son of Joseph, (as was supposed). Who supposed? James the Lords brother? He never mentioned it and he was the Lord's flesh brother. Neither did Jude in his epistle, who was also the Lord's flesh brother. Don't tell me his own flesh and blood brothers didn't know about the virgin birth Who was it that "supposed"? These disciples came from Capurnium only a few miles from Nazareth, where Jesus grew up. Were these men left in the dark about the amazing events of Jesus' birth? It is very difficult to keep any thing secret in a small town. What a great sign this truly is, yet not one of the apostles seemed to know anything about it, at least if they did, they never mentioned it in any of their epistles. **<u>Not once!</u>** When they witnessed to the Lord's sonship they spoke of his <u>resurrection</u> and <u>his lineage to David</u>. 1st Timothy 1 : 4, Paul says: **Neither give heed to endless genealogies which only render questions rather than godly edifying which is in faith.** Whose genealogy was Paul talking about? Peter's? John's? Of course not! It was <u>Jesus' lineage back to David</u> that Paul was speaking of. They did not question that Jesus was the son of David through Joseph. They questioned whether his lineage went back through the evil idol worshipping kings or through the lineage found in Luke or perhaps some other way we know nothing about. We do know that matthew's genealogy was different from Luke's. The bible says this linage would be THROUGH HIS SONS and that's the way it is in both Matthew and Luke. The words, **"it was supposed"**, were added by the translators.

You may wonder; why the lineage back to the Levites was never mentioned by the apostles as their witness to the sonship of Jesus. The reason is explained in Isaiah chapter 6 starting with verse9: And he said, **Go, and tell this people, Hear ye indeed, but understand not; and see ye indeed, but perceive not. Make the heart of this people fat, and make their ears heavy, and shut their eyes; lest they see with their eyes, and hear with their ears, and understand with their heart, and convert, and be healed.** The Jews of Jesus's time did not know that the savior had to come the first time and establish the Melchizedek priesthood. An unblemished Lamb had to be sacrificed for this. However the eyes of the Jews were closed to this because of the POWER of the word of God. God said they would not know him and they didn't. They believed that he was coming to establish a kingdom and deliver them from the power of Rome. This was because of the traditional teaching in the synagogs that the messiah was going to establish a kingdom and rule the world in Jerusalem. But only one Jew, in Jesus's time, ever knew that the Lord had come to be a sacrifice instead of a king. That was Jesus himself. How do we know this? James and John asked to be placed on his right hand in his kingdom. They had seen his mighty works, they thought his kingdom would be set up right then. A large following of people were on hand when he rode triumphant into Jerusalem on the ass. Two weeks earlier they had seen him raise Lasurus from the dead. They thought he was going into Jerusalem to proclaim his kingdom. John The Baptist faced death in Herrod's dungeon and like any man facing death he clung to hope. His hope was Jesus. He too, thought that Jesus had come to establish a kingdom. That's why he sent to Jesus to ask: **"Are you the one to come or do we look for some one else"?** You know Jesus's answer; The dead are raised, the blind see, and the kingdom is preached to the poor. Did this answer cause John to know? I cannot say. Only once did Jesus come right out and tell one of his fellow Jews just who he was. This was the man who had been born blind. Jesus healed him and because he made the pharisees mad, they cast him out of their system. He was no longer a Jew he was cast out, therefore, Jesus did not go against his Father's word by telling him that he was the son of God. This account is found in John chapter 9. He also came right out and told the woman at the well that he was the Son of God but she was a Samaritan gentile not a Jew.

Another reason why I know that the Jews did not know Jesus had come to suffer in order to establish an everlasting priesthood, was the account of two men on their way to Emmamus. Luke 24 :13-27 says; **And, behold, two of them went that same day to a village called Emmaus, which was from Jerusalem about threescore furlongs. And they talked together of all these things which had happened. And it came to pass, that, while they communed together and reasoned, Jesus himself drew near, and went with them.16 But their eyes were holden that they should not know him. And he said unto them, What manner of communications are these that ye have one to another, as ye walk, and are sad? And the one of them, whose name was Cleopas, answering said unto him, Art thou only a stranger in Jerusalem, and hast not known the things which are come to pass there in these days? And he said unto them, What things? And they said unto him, Concerning Jesus of Nazareth, which was a prophet mighty in deed and word before God and all the people: And how the chief priests and our rulers delivered him to be condemned to death, and have crucified him. But we trusted that it had been he which should have redeemed Israel: and beside all this, to day is the third day since these things were done. Yea, and certain women also of our company** made us astonished, which were early at the sepulchre; And when they found not his body, they came, saying, that they had also seen a vision of angels, which said that he was alive. And certain of them which were with us went to the sepulchre, and found it even so as the women had said: but him they saw not. Then he said unto them, O fools, and slow of heart to believe all that the prophets have spoken: Ought not Christ to have suffered these things, and to enter into his glory? And beginning at Moses and all the prophets, he expounded unto them in all the scriptures the things concerning himself.

The church received the message through these men that Jesus had come to suffer many things at the hands of the unbelieving Jews. The Jews did not know until after the resurrection and the new covenant was instituted that Jesus had come to be a Lamb not a king. They still, to this day, deny the following scriptures found in Isaiah 53:6: **All we like sheep have gone astray; we have turned every one to his own way; and the Lord hath laid on him the iniquity of us all. He was oppressed, and he was afflicted, yet he opened not his mouth: he is brought as a lamb to the slaughter, and as a sheep before her shearers is dumb, so he opened not his mouth. He was taken from prison and from judgment: and who shall declare his generation? For he was cut off out of the land of the living: for the transgression of my people was he stricken. And he made his grave with the**

wicked, and with the rich in his death; because he had done no violence, neither was any deceit in his mouth. Yet it pleased the Lord to bruise him; he hath put him to grief: when thou shalt make his soul an offering for sin, he shall see his seed, he shall prolong his days, and the pleasure of the Lord shall prosper in his hand. Now you truly see the power in the word of God. It must come to pass. They had blinded themselves by their traditional teaching about a kingdom. When God spoke through Isaiah that they would be blind to the purpose of Jesus's first coming. It happened exactly that way. The rabbi's of his day knew only of a great king who would come some day and establish an eternal kingdom. But they were totally blinded to the scripture that he must come first to suffer so that the eternal priesthood could be established. This is why they knew nothing of the covenant that God made with Phinehas the Levite. Without our great High Priest to intercede for us with the Father there would be no Body of Christ.

Thus we see that God made seed covenants with men in the old testament. We know that **only flesh** has SEED. This is the law of nature that God himself placed there. I am not saying that God could not overrule the laws that he made but in this case he did not because we needed a **legal** redeemer. A man made like us in **all** things was our redeemer. Almighty God never half does anything, Jesus redeemed us as a man that was made like us in all things. He was not the product of a pagan lie. He was **the word**. It is not the nature of spirit to have seed, for Jesus himself said, that in the resurrection, there is neither male nor female but all will be as the angels of God. God does not go against his own word.

There were many more prophecies in the bible concerning these seed covenants that must be brought out. Just in case you still think that Solomon was the one that God was talking about who would build the house of David, there is a prophesy found in Jeremiah who lived many centuries after Solomon. This prophesy confirms that it was not Solomon who would build this house that God was speaking about. In Jeremiah's time the great temple had already been built. In fact, it was in Jeremiah's time that it was torn down. However, God said it would be an eternal house that he would build. The word of the Lord came to Jeremiah just before the people of Israel were taken into captivity in Babylon concerning his covenant with David. The Lord God wanted to make sure these seed covenants would not be forgotten. Jeremiah chapt. 33 vs. 17,18 **Thus saith the Lord, David shall never want a man to sit upon the throne of the house of Israel; neither shall the priests the Levites want a man before me to offer burnt offerings, and to do sacrifice continually.** Also verse : 20, 21 **Again the word of the Lord came to Jeremiah, saying, Thus saith the Lord; If you can break my covenant of the day, and my covenant of the**

night, that there should not be day and night in their season; then may also my covenant be broken with <u>David my servant</u> that he should not have a son to reign upon his throne, and with the <u>Levites my priests,</u> my ministers. Notice that both David and the levites were mentioned in these prophesies. This everlasting Melchezdek priesthood was established by Jesus. He had to be of the seed of the Levites as well as the Jews. The Levites were types and shadows of the everlasting priesthood and seed of David was a type and shadow of the everlasting kingdom. Jehoiachin, who was taken captive into Babylon, was the last recorded king of Israel. This prophesy still speaks of David"s covenant for an everlasting kingdom of his Son. God mentions his covenant with the Levites as well. This is sometimes overlooked entirely. It was overlooked by the early church as well as the Jews of Jesus's time,. This, of course, is not the only place in the old testament that the seed covenants are confirmed.

Note also the 89th Psalm: **I have made a covenant with my chosen, I have sworn to David my servant, thy SEED will I establish forever and build up thy throne to all generations.** Psalm 106 vs. 29-31: **They provoked him to anger with their inventions and the plague brake on them. Then stood up Phinehas and executed judgment and so the plague was stopped and that was counted to him for righteousness and to all generations for evermore.** Also Malachi chapter 2 vs. 4,

5 : And you shall know that I have sent this commandment unto you, that my covenant might be with the Levites says the Lord of hosts. My covenant was with him of life and peace; (numbers 25;12) and I gave them to him because he feared me and was afraid before my name. You see, these seed covenants were witnessed to by the prophets long after they were made so that they would not be forgotten, **Jesus was the <u>seed</u>** of these men with whom God made these covenants. Jesus was flesh just like all men are flesh. But this seed did something no other flesh has ever done, He lived 33 years without sin. What a miracle this was! He also ascended to the Father in the flesh. He stood before the face of the Father, and even now it stands before him to intercede for us. Isaiah once stood before him and he fell on his face and cried out**, woe is me for I am undone, depart from me lest I die!** No flesh can stand before him and live because he has said the soul that sins, it shall die. The power of his word is so strong that flesh would die in his presence. Jesus is our great high priest. he has made the oblation, the sacrifice for the sins of the people once and for all and is now our intercessor forever. Praise God!

We have seen what the prophets have spoken under the influence of the Holy Ghost concerning Jesus, the word, now we shall look into the New Testament to see if the apostles mentioned the virgin birth in their witness to others about Jesus the Son of God. Go to acts chapter 1 vs. 22: Judas had committed suicide, so the rest were preparing to select another to take his place. It was written in Psalms 69: 25 and Psalm 109: 8; **Let his habitation be desolate and let no man dwell therein: and his bishopric another take."** In Acts 1:22; it says: **Beginning from his baptism by John unto that same day that he was taken up from us, someone must be ordained to be a witness with us of his <u>resurrection.</u> (the <u>ONLY</u> sign given to this generation was <u>the resurrection)</u>** By this we see that the apostles were ordained to confirm that Jesus was indeed THAT PROPHET who was promised to Moses. Their witness was based on the resurrection. THE VIRGIN BIRTH WAS NOT MENTIONED. Jesus also gave a witness to his sonship in Matthew 12:39: **An evil and adulterous generation seeks after a sign, and there shall be no sign given to it (only one) but the sign of the prophet Jonas: for as Jonas was 3 days and 3 nights in the whale's belly; so shall the son of man be three days and three nights in the heart of the earth (the tomb).** Jesus himself said that this generation would receive only <u>ONE</u> sign, not two; and this sign is the **<u>resurrection,</u>** not a virgin birth. How many people hold the sign of the virgin birth up higher in their minds than the resurrection? Yet, as far as Jesus was concerned there was only one sign given to our generation and that was the resurrection not the virgin birth. And as we have said not one of the apostles ever mentioned the virgin birth as a sign. The sign to the old generation given by Paul, Peter, James and John was his linage to david. The sign given to the new generation was the resurrection from the dead. Though many people say that the virgin birth is not a sign to them this writer finds it hard to believe that theyreally mean this. The resurrection is the sign that the apostles always witnessed to, not a virgin birth. Please note that he said that he would be in the tomb 3 days and 3 nights. This is three 24 hour periods; a total of 72 hours. How then can there be a "Good Friday" if he rose from the dead on Sunday morning? This is not 72 hours. Have the scriptures been altered on this?

Go back to Acts chapter 2 vs. 24-36: Read it yourselves; take no man's word. Go the bible yourself to see what God has said. Peter witnessed to <u>TWO THINGS</u>: First, <u>THE RESURRECTION,</u> 2nd his <u>LINEAGE TO DAVID,</u> nothing else! Peter never mentioned a virgin birth. He also witnessed to the resurrection in Acts 3: 15; 4: 2; 4: 33; 5: 30. There are many other places found in the book of Acts, in which the apostles witnessed to the resurrection of Jesus Christ in order to prove his sonship to their fellow Jews and the gentiles. Yet not one word was mentioned about the virgin birth! The absence of any account of the virgin birth is scriptural evidence that there was no virgin birth.

Now let's see what the apostle Paul used as the witness to confirm the sonship of Jesus. Paul lived in the same time period as the apostles. Paul knew them personally. He mentioned Peter , Mark, James ,John and other men who had been with Jesus Christ.

123

Let's see how he opened the book of Romans. Romans chapter 1 vs. 1-4; **Paul a servant of Jesus Christ, called to be an apostle, separated unto the gospel of God, which he had promised before by his prophets in the holy scriptures. Concerning his son Jesus Christ our Lord, which was made of the seed of David according to the flesh; and declared to be the son of God with power, according to the spirit of holiness, by the resurrection from the dead.** Here again, Paul also witnesses to TWO things, the resurrection from the dead, and his lineage to David by being his flesh SEED. Again, no mention of a virgin birth. Doesn't it seem strange that Paul who knew personally the writers of the gospels, Matthew, Mark, and Luke never mentioned such a phenomenal sign as the virgin birth? Unless, of course, there was no virgin birth. This writer believes it was added many years later by pagans who wanted political power so they could bring in pagan "converts". The Roman emperors claimed divinity by saying they were procreated by the Roman gods, so they would be worshiped by the Roman people. Paul preached one of the greatest universal, ecumenical sermons that has ever been preached when he preached to the intellectuals on Mars Hill. Yet, in his sermon to them, he witness to the resurrection of Jesus Christ from the dead not a virgin birth. They would have laughed him to scorn if he had preached a virgin birth. Instead, many of them were so impressed by the news of the resurrection that they wanted to hear more.

Would they have listened if he had claimed Jesus's divinity by giving a sign of a virgin birth?

In Acts chapter 13 Paul preached a sermon in the synagogue at Antioch to the Jews. Here again he witnessed to the sonship of Jesus by the testimony of the <u>lineage to David</u> and the <u>resurrection.</u> Again, no mention of the virgin birth. Romans chapter 2: 16-17: **For very surely he took not on him the nature of angels but he took on him the seed of Abraham. Wherefore in ALL THINGS it behoved him to be made like unto his brethren, that he might be a merciful and faithful high priest in things pertaining to God, to make reconciliation for the sins of the people**. Paul says here that he was made like men in ALL things. In this case does ALL mean all inclusive? Someday, Jesus will judge all flesh because he said in the gospel of John that all judgment is given to the SON OF MAN. He lived 33 years on this earth without sin. This is a much greater miracle than a virgin birth. In 2nd Timothy Chapter 2vs. 8 Paul writes: Remember that Jesus Christ of **the seed of David** was **<u>raised from the dead</u>** according to my gospel. Once again Paul emphasizes **the linage to David** and the **resurrection,** yet , not one word about a virgin birth.

There is another scripture that people use to prove that Jesus was of a virgin birth; Luke chapter 20 VS. 41-44; also, Mark chapter 12 VS. 35-37; **How say the scribes that Christ is the son of David? David himself said by the Holy Ghost, the Lord said**

to my Lord, sit on my right hand until I make your enemies thy footstool. David himself called him Lord how then can he be his son? Jesus was being tested by the scribes. They had been asking him questions concerning the law in order to trick him and make him show up as a deceiver of the people. However, he answered all their questions with such wisdom that they dared not ask him any thing else because he was making them look bad in the eyes of the people. Keep in mind that the scribes and pharisee were trying to find an excuse to kill him. Why? Because they thought he was an imposter. They were like many religious leaders of today; in their own eyes they are righteous and ready for Jesus to return. But the scriptures say that when he comes the second time that many of these preachers will say: **"Have we not prophesied in thy name, and in thy name cast out devils, and in thy name done many wonderful works? Then he will profess to them depart from me you workers of iniquity for I never knew you**. (Matthew 7; 22). The scribes and pharisees were expecting "that prophet" to come because it was prophesied in the book of Daniel, chapter 9 vs. 25. All the Jews knew the time was at hand. They expected Christ to come as a great king, high and lifted up. He would show great signs and wonders and lift them up before the people for their religious piety. Instead he came as a poor man preaching the gospel to the poor people and worse yet, he called them snakes and hypocrites. You can believe that they had already checked his lineage to see if he actually was of the seed of David. In their minds they did not question this. They knew that he was the son of Joseph because they knew his father and mother.(John 1: 45; 6: 42;) They had already questioned his right to say that he was "that prophet" because they thought he was from Galilee not Bethleham. Jesus was a poor man, a nobody. He didn't come with great pomp showing signs and wonders. He called them religious bigots, and they hated him. This is the reason that Jesus made the above statement: **How say the scribes that Christ is the son of David?** He was saying that they knew the messiah had to be the son of David yet he was denied because he was a poor man. He knew that they were denying his Lordship on the basis of being a poor man like the rest of those that he ministered to. He loved the poor just as much as he did the rich and he favored none. Jesus was "rocking the boat" and exposing them as the evil, greedy, worldly, people that they actually were. He told them they were like cups washed on the outside but on the inside they were filthy. Jesus didn't mince words with these rulers of Israel he told it like it was. That's why they eventually killed him. They were men of power not pantywaists and neither was Jesus!

When we go back into ancient history we see the Mother-Child figure used thousands of years before Jesus was born. In the book, "The Two Babylons", by Hislop he states that the Babylonians had in their most popular religion a Mother goddess and son image. From Babylon this mother- child figure spread to the ends of the Earth. In Egypt this mother – son image was worshiped as Isis and Osiris. In India, even to this day, it is worshiped as Isi and Iswara. In Asia, she was Cybele and Deoius. In Greece, she was Ceres the goddess of peace with Plutus in her arms. Even in Tibet, China and Japan the Jesuit missionaries were astonished to find the counter part of the Madonna and Child as devoutly worshiped as much as it was in Rome itself. In China she was pictured with the child in her arms and with Glory

The above illustrations are from Hislop's book, "The Two Babylons"

around her exactly as if a Roman Catholic priest had done the art work. Yet this image was worshiped many centuries before Jesus Christ was born. This certainly makes one believe that this virgin birth idea existed long before the birth of Christ. Therefore, it is entirely possible that the doctrine of the Virgin Birth came into the church long after the apostles were dead. There was a Virgin Mother and Child image found in Germany, England, and in Scandinavia as well.

Hislop says in his book " The Two Babylons" that the Virgin Mary worshiped by the Roman Catholic Church is the same as that worshiped in ancient Babylon. He also says that Christmas never came about until the 3rd century and didn't become popular until the 4th century.

In the book, "The Outline of History", written by H. G. Wells he says the people found it impossible to believe that Buddha was the son of a mortal (flesh) father, therefore the saying was that he was miraculously conceived through his mother's dream of a beautiful white elephant.

Illustrationsfrom H.G.Wells' book "Outlie of History"

We can surmise from all this that the idea of a miraculous conception was very common through out ancient history in most all the pagan religions. Since none of the apostles never mentioned a virgin birth in any of their epistles to the church, we can say that the virgin birth was introduced into the Roman church long after their death. The pictures we have here are of the various pagan religions long before Jesus Christ was born. They Show the mother child figure as they appeared in these pagan religions. Notice the simularity to the pictures we see of Mary and Jesus. I hope these things that have been documented by historians in ancient history help you to see that there is no credibility in the account of the virgin birth. We have seen that none of the apostles ever mentioned a virgin birth in their epistles to the early churches. If the virgin birth was known by the apostles it would certainly seem likely that they would have mentioned it in their letters to the churches.

Some of you may consider me a heretic who is changing the bible to fit my own ideas. However, everything that I have written is backed up by scriptural references except for one thing. That exception is the statement that Mary's father was a Levitical priest. This fact is not based upon scripture it is only based upon historical evidence. We do know that Elizabeth, John the Baptist's mother, was of the daughters of Aaron. Mary was her cousin. This does establish a scriptural connection to Aaron. Phinehas was Aaron's grandson. Elizabeth and Mary were cousins. Faustus a learned historian at the time of the early church proclaimed that Mary's father was of the Aaron priesthood. Notwithstanding, there is one thing that I do know, night and day still come at their appointed times, therefore, the promise made to Phinehas for the everlasting priesthood through his SEED has been kept. Who else could have kept it but Jesus? I believe it was kept through Mary the mother of Jesus. I also know, by the scriptures, that the seed covenant for an everlasting kingdom through the seed of David has also been kept through Joseph the father of Jesus. This covenant was kept through the seed of David's sons just as God said it would be when he spoke to Nathan the prophet. Joseph the father of Jesus was of this son seed. He put not upon him the likeness of angels but was made like men in ALL things.

Jesus's not being of a virgin birth does not detract from his being God manifested in the flesh. But if I thought for one instant that all that was spoken of him by the Law and the prophets did not take place exactly as they predicted, then I could not believe upon him as word. However, he was the word. For this reason, there can be no discrepancies about him.

The fact that I find no discrepancy nor any fault in him or the words he spoke makes me cry out to him "My Lord and my God." You can choose what you want to believe, for Jesus says "He who has ears to hear let him hear". He told peter to feed his sheep, He didn't say, kill all those who don't believe like you do. I believe his sheep will hear his voice!. I am in no way pushing the paganism or the Hollywood version of Jesus Christ. These Images of Jesus are ridiculous. But I am saying that he was surely God in the flesh. And he did not just speak the truth, **HE WAS THE TRUTH!** In order to be the truth there can be no discrepancies about him. He called himself "the son of man" not "the son of woman".

On that day when men are standing in judgment there will probably be those who will try to justify themselves by saying to Jesus, "You had a special birth which was directly of the Father but we are just ordinary men how can you condemn us?" But the words of Paul, who said, "He put not on him the likeness of angels but was made like men in all things" will be spoken to them. We had a legal redeemer he was not half spirit and half man. He redeemed us because he was a man just like all men, yet, he was without sin. Which is the greater miracle, a virgin birth or 33 years without a single sin? Many judges pass death sentences on men in our court systems every day, but these judges don't give their lives too, do they? However, God put on flesh and became a man, and died his own sentence of death that we might be free of death forever. We had a **legal** redeemer, a man made like us in **all things**. Now, the holy Ghost can enter into this world legally to live in the redeemed of Jesus. Many say he could not have redeemed us if he had been tainted with Adams SEED. But his mother Mary was of the seed of Adam, why didn't this taint him? The devil won over mankind by enticing Adam to sin. But because Jesus was a man made like all the rest of us yet he was not blemished with sin, we can now enter into the kingdom of God through HIM.

A man,(God in the flesh) has set us free from the law of sin and death. Jesus was of the seed of Levi, the priests of God. He was also of the seed of David the King of Israel. These were two prototypes that signified the everlasting Priesthood and the everlasting Kingdom.

His first coming was to establish the Melchezdek priesthood by the sacrifice of his unblemished SELF. The Jews were blinded to this fact. His second coming will be as a King to establish the everlasting Kingdom of God. My prayer for you is that God will reveal his precious son Jesus Christ to you. And, in that day you will see him as he is. AMEN

About The Author

The author of this book is a veteran, a college graduate, and a fundamental Christian. His background is in science. He has been a Christian for about 50 years. He has been writing Christian tracts for about 25 years. He and his wife have their own website with Christian tracts on it. Because of this author's science and mathematical background he has been shown many "types and shadows" of Jesus and God's purposes in him by the blessed Holy Ghost.

www.ingramcontent.com/pod-product-compliance
Lightning Source LLC
Chambersburg PA
CBHW021957170526
45157CB00003B/1024

* 9 7 8 1 4 1 8 4 3 1 2 2 8 *